1200/N

ULTRASONICS

Theory and Application

ULTRASONICS
Theory and Application

G. L. GOOBERMAN
B.Sc., Ph.D.
Lecturer in Electrical Engineering
University of Salford, Lancashire

HART PUBLISHING COMPANY, INC.
NEW YORK CITY

Editor's Foreword

Ultrasonics deals with sound waves which are at such a high frequency that they cannot be detected by human ears. This is not a very rigid definition, because the frequency above which sound cannot be heard varies from individual to individual, and within the lifetime of any given individual. Generally speaking, as we grow older the cut-off frequency is reduced. This means that by the time we have worked long enough to be able to afford really good hi-fi equipment, we tend to be no longer young enough to hear the high notes it is capable of producing. That is but a particular example of a powerful general law well known to experimental scientists and engineers all over the world.

Ultrasonics is a rewarding study both from the scientific and the engineering points of view. From the scientific point of view, the subject embraces a study of wavemotion, including radiation. It includes the effect on the molecules of the materials through which the energy passes. From the engineering point of view, ultrasonics brings in a study of transducers of various kinds, and a study of measurement techniques. Ultrasonics is also finding increasing application in various fields, such as quality control in production engineering, and for medical application, particularly where the more conventional X-ray techniques are undesirable for some special reason.

It is hoped that this book will be useful to those interested in ultrasonics either from the point of view of the appreciation of the basic physics involved or from the point of view of their utilisation in engineering applications.

J. H. CALDERWOOD

Preface

This book has been written as an introduction to the properties and applications of ultrasonic waves. Since it starts from first principles it is suitable for undergraduates pursuing a course in ultrasonics and also for research workers who wish to use ultrasonic techniques but who have little previous knowledge of the subject. The mathematical skill required of the reader has been kept to a minimum but a slight knowledge of electrical circuit theory and elementary thermodynamics is assumed. Not all the steps in the many mathematical derivations are given in detail but in most cases the steps are described in such a way that the reader can, if he so wishes, insert them for himself. So as not to overload with references only the more pertinent are given.

The first two chapters cover the necessary theory of wavemotion and radiation in some detail. The next two chapters cover transducers, the transmission line equivalent circuits being derived as well as the more usual simple tuned circuit type of equivalent circuit. Chapter 5 deals with the effects of high power ultrasonic radiation including a comprehensive survey of cavitation. Chapter 6, probably the most complicated one in this book, consists of a general treatment of relaxation. Its effect on the dispersion and absorption of ultrasonic waves is given, and in the following two chapters experimental techniques and their results for longitudinal waves are described. Shear waves are dealt with in chapter 9, and chapter 10 covers propagation in solids including some of the recent work on microwave ultrasonics. Chapter 11 deals with a few of the more important industrial and medical applications of low power ultrasonics. The appendices contain a selection of useful ultrasonic data plus a short bibliography of books suitable for further reading.

In a book of this size it is not possible to cover all aspects of ultrasonics either in breadth or depth and therefore the topics dealt with must necessarily represent a selection made at the personal whim of the author. However, the topics covered should be sufficient in number, and the treatment adequate, to enable those readers who so wish to go on to the more detailed treatises with an adequate background knowledge.

G. L. GOOBERMAN

Contents

1
General Principles of Wavemotion

1.1 Wavemotion

The term wavemotion is usually taken to mean the transmission of a disturbance through a medium in such a way that energy is propagated but that after the passage of the disturbance the medium is in the same state as it was before the arrival of the disturbance. This definition can be faulted in the case, for example, where the passage of a shock wave strains the medium beyond its elastic limit. However, this definition suffices for our present purposes.

We will consider a disturbance, of unspecified nature, which can be measured by some parameter θ and which travels without change of either magnitude or shape along the x axis of a co-ordinate system with a velocity c. At a time $t = 0$ the parameter θ will be some function of x which we will denote by $f(x)$ so that we can write:

$$\theta = f(x) . \qquad . \quad . \quad . \quad . \quad . \quad \textbf{(1.1)}$$

After a time t the disturbance will have travelled a distance ct. As we assume that the disturbance travels without change in either magnitude or shape θ will still be given by equation (1.1) provided that we move the origin of our co-ordinate system to the position given by $x = ct$. Thus in terms of our original co-ordinate system we have:

$$\theta = f(x - ct) . \qquad . \quad . \quad . \quad . \quad \textbf{(1.2)}$$

For a disturbance travelling in the opposite direction we have:

$$\theta = f(x + ct) . \qquad . \quad . \quad . \quad . \quad \textbf{(1.3)}$$

If θ, as given by either equation (1.2) or (1.3), is differentiated twice with respect to x and then to t we obtain the following differential equation:

$$\frac{\partial^2 \theta}{\partial t^2} = c^2 \frac{\partial^2 \theta}{\partial x^2} \quad . \qquad . \quad . \quad . \quad . \quad \textbf{(1.4)}$$

Equation (1.4) is the fundamental equation of wavemotion and whenever an equation of this form appears we can be sure that we are dealing with wavemotion of some sort.

So far we have not specified the type of disturbance which is being propagated. In ultrasonics we are concerned with the properties of elastic waves in solids, liquids and gases.

1.2 Elastic Wavemotion

In order to derive the equation of wavemotion for elastic waves in a medium we will consider a small element of the medium, of mass m, under the influence of mechanical forces. If these forces are unbalanced the small element will be accelerated in, say, the x direction with an acceleration given by $\frac{\partial^2 \xi_x}{\partial t^2}$ where ξ_x denotes the displacement of the

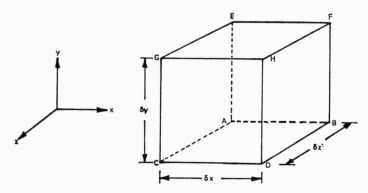

Fig. 1.1 Rectangular parallelepiped used for stress analysis.

element in the x direction. If the net force acting on the element be F, then by Newton's second law of motion we have:

$$\frac{\partial^2 \xi_x}{\partial t^2} = \frac{1}{m}\,F \quad . \quad . \quad . \quad . \quad . \quad (1.5)$$

If we can express F as $\frac{A \partial^2 \xi_x}{\partial x^2}$, where A is a constant, we have an equation of the same form as equation (1.4) in which θ now represents the displacement of the element; that is we will have derived the equation of wavemotion for elastic waves. The force F can be obtained in terms of the displacement from the stress–strain relations for the medium and these are derived in the next section.

1.3 Stress–Strain Relationships

The most general form of the stress–strain equations for a medium are those appertaining to a solid; those for liquids and gases being simplified forms of these since, unlike solids, liquids and gases cannot usually support a shear. This last statement will require some modification later on when we deal with very high frequency waves.

We consider an infinitesimal rectangular parallelepiped, as shown in Fig. 1.1, situated somewhere within a solid medium, with sides of length δx, δy and δz which are parallel to a set of rectangular axes x, y and z. When this medium is stressed the parallelepiped will, in general, be displaced from its original position and simultaneously be deformed. The most general displacement can be regarded as the vector sum of

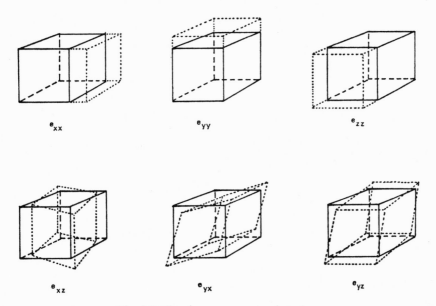

Fig. 1.2 The six components of strain.

three displacements, each of which is parallel to one of the axes, plus a rotation, while the deformation can be regarded as being compounded of changes in length of each side of the parallelepiped allied with a shear. We require to find equations relating the stresses acting on the parallelepiped to the resulting strains.

We consider first the strain and we denote each strain component by a term of the type e_{ij} in which each of the subscripts can be x, y or z and which denote the directions in which the strain occurs. There are six such strain components and these are illustrated in Fig. 1.2. When $i = j$ we have a longitudinal or tensile strain and when $i \neq j$ we have a shear strain. If we let the displacements of point A of the parallelepiped under the influence of the stresses acting on the medium be ξ_x, ξ_y and ξ_z parallel to the x, y and z axes respectively then the displacements of point H are given by $\xi_x + \dfrac{\partial \xi_x}{\partial x}\delta x$, $\xi_y + \dfrac{\partial \xi_y}{\partial y}\delta y$ and $\xi_z + \dfrac{\partial \xi_z}{\partial z}\delta z$. Since the tensile strain in a given direction is defined as the proportional change in length in that direction we have:

$$e_{xx} = \frac{\partial \xi_x}{\partial x}$$

$$e_{yy} = \frac{\partial \xi_y}{\partial y} \quad . \quad . \quad . \quad . \quad . \quad . \quad . \quad \textbf{(1.6)}$$

$$e_{zz} = \frac{\partial \xi_z}{\partial z}$$

In order to find expressions for the shear strains we consider the plane *GHCD* of our parallelepiped and suppose it to be sheared as shown in Fig. 1.3. Shear strain is defined as the change in the angle between two

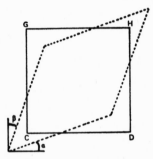

Fig. 1.3 Shear strain applied to plane *GHCD* of Fig. 1.1.

sides of the rectangle, that is as the angle $\alpha + \beta$. Now $\tan \alpha = \dfrac{\partial \xi_y}{\partial x}$ and

and $\tan \beta = \dfrac{\partial \xi_x}{\partial y}$ and if we restrict our analysis to infinitesimal strains these angles, α, β, will be very small so that their tangents are practically equal to the angles. Thus we have:

$$e_{xy} = \frac{\partial \xi_y}{\partial x} + \frac{\partial \xi_x}{\partial y} \quad . \quad . \quad . \quad . \quad \textbf{(1.7a)}$$

Similarly for the other planes we have:

$$e_{zx} = \frac{\partial \xi_x}{\partial z} + \frac{\partial \xi_z}{\partial x}$$

$$e_{yz} = \frac{\partial \xi_z}{\partial y} + \frac{\partial \xi_y}{\partial z} \quad . \quad . \quad . \quad . \quad \textbf{(1.7b)}$$

Since α does not necessarily equal β the diagonal can rotate when the parallelepiped is deformed. In the example given in Fig. 1.3 the rotation is $\frac{1}{2}(\alpha - \beta)$ so that denoting the rotation about the z axis by w_z we have:

$$w_z = \tfrac{1}{2}(\alpha - \beta) = \tfrac{1}{2}\left(\frac{\partial \xi_y}{\partial x} - \frac{\partial \xi_x}{\partial y}\right) \quad . \quad . \quad . \quad \textbf{(1.8a)}$$

Similarly for the rotations about the x and y axes, denoted by w_x and w_y respectively, we have:

$$w_x = \tfrac{1}{2}\left(\frac{\partial \xi_z}{\partial y} - \frac{\partial \xi_y}{\partial z}\right)$$

$$w_y = \tfrac{1}{2}\left(\frac{\partial \xi_x}{\partial z} - \frac{\partial \xi_z}{\partial x}\right) \quad . \quad . \quad . \quad . \quad \textbf{(1.8b)}$$

We now require expressions for the stresses acting on the sides of the parallelepiped. At any point the stress can be resolved into three components parallel to the x, y and z axes. We denote each stress by σ_{ij}, which is taken to be positive for a tension, where the first subscript

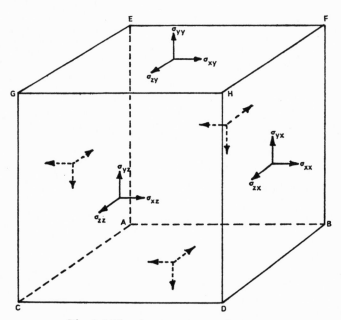

Fig. 1.4 The six components of stress.

denotes the direction in which the stress acts and the second the plane on which it is acting, with the plane designated by the direction in which its normal points. These stress components are illustrated in Fig. 1.4. Since the parallelepiped is in equilibrium, the tensile stresses parallel to any one axis must be equal, otherwise the element would accelerate. Thus we can have only three independent tensile stresses. Similarly the shear stresses must also balance otherwise the parallelepiped would rotate continuously so that we have only three independent stresses of the type denoted by σ_{ij} in which $i \neq j$. Thus we find that the stress acting on the parallelepiped can be completely represented by six independent components; the three tensile stresses σ_{xx}, σ_{yy} and σ_{zz} plus the three shear stresses σ_{xy}, σ_{yz} and σ_{xz}.

We now have to relate the strains to the stresses. Experimentally we know that for small stresses, well below the elastic limit of the medium, strain is directly proportional to stress. We can generalise this by writing each stress component at a given point as a linear function of the six strain components. Thus we have:

$$\left. \begin{aligned}
\sigma_{xx} &= c_{11}e_{xx} + c_{12}e_{yy} + c_{13}e_{zz} + c_{14}e_{yz} + c_{15}e_{zx} + c_{16}e_{xy} \\
\sigma_{yy} &= c_{21}e_{xx} + c_{22}e_{yy} + c_{23}e_{zz} + c_{24}e_{yz} + c_{25}e_{zx} + c_{26}e_{xy} \\
\sigma_{zz} &= c_{31}e_{xx} + c_{32}e_{yy} + c_{33}e_{zz} + c_{34}e_{yz} + c_{35}e_{zx} + c_{36}e_{xy} \\
\sigma_{yz} &= c_{41}e_{xx} + c_{42}e_{yy} + c_{43}e_{zz} + c_{44}e_{yz} + c_{45}e_{zx} + c_{46}e_{xy} \\
\sigma_{zx} &= c_{51}e_{xx} + c_{52}e_{yy} + c_{53}e_{zz} + c_{54}e_{yz} + c_{55}e_{zx} + c_{56}e_{xy} \\
\sigma_{xy} &= c_{61}e_{xx} + c_{62}e_{yy} + c_{63}e_{zz} + c_{64}e_{yz} + c_{65}e_{zx} + c_{66}e_{xy}
\end{aligned} \right\} \quad (1.9)$$

where the coefficients of form c_{rs}, with r and s ranging from 1 to 6, are the elastic constants of the material. These equations are not directly verifiable but they lead to verifiable mathematical conclusions.

It would appear from equations (1.9) that thirty-six coefficients are necessary in order to completely describe the behaviour of a medium. However, it can be shown[1.1] that $c_{rs} = c_{sr}$ so that the thirty-six coefficients are reduced to twenty-one. If the medium possesses planes of symmetry the number of coefficients is further reduced until, for a completely isotropic medium, they number two only. For this last case the two coefficients are usually denoted by λ and μ and are known as Lamé's constants. Lamé's constants are related to the elastic constants by the following equations[1.2]:

$$\begin{aligned}
\lambda &= c_{rs} \ (r \neq s) \\
\mu &= c_{44} = c_{55} = c_{66} \quad . \quad . \quad . \quad . \quad (1.10) \\
\lambda + 2\mu &= c_{11} = c_{22} = c_{33}
\end{aligned}$$

Thus the final stress–strain equations for an isotropic medium become:

$$\left. \begin{aligned}
\sigma_{xx} &= (\lambda + 2\mu)e_{xx} + \lambda e_{yy} + \lambda e_{zz} \\
\sigma_{yy} &= \lambda e_{xx} + (\lambda + 2\mu)e_{yy} + \lambda e_{zz} \\
\sigma_{zz} &= \lambda e_{xx} + \lambda e_{yy} + (\lambda + 2\mu)e_{zz} \\
\sigma_{yz} &= \mu e_{yz} \\
\sigma_{zx} &= \mu e_{zx} \\
\sigma_{xy} &= \mu e_{xy}
\end{aligned} \right\} \quad . \quad . \quad (1.11)$$

In practice four elastic constants, viz. Young's modulus, bulk modulus, rigidity modulus and Poisson's ratio are used rather than the two Lamé constants.

Young's modulus E, which is defined as the ratio of stress to strain for a prismatic specimen which is uniformly stressed longitudinally and which is free at its lateral surfaces, is given by the ratio σ_{xx}/e_{xx} with the remaining stresses equated to zero.

The bulk modulus k, which is defined as the ratio of the applied pressure to the fractional volume change when the specimen is subjected to a uniform hydrostatic pressure, is given by $-\sigma_{ii}/(e_{xx} + e_{yy} + e_{zz})$.

The rigidity or shear modulus μ, which is defined as the ratio of the shear stress to the shear strain, is given by the ratio σ_{ii}/e_{ij} ($i \neq j$) and is simply Lamé's second constant μ.

Poisson's ratio ν, which is defined as the ratio between the lateral contraction and the longitudinal extension of the specimen with its lateral surfaces free, is given by the ratio $- e_{yy}/e_{xx}$.

Thus we have[1.3]:

$$E = \frac{\mu(3\lambda + 2\mu)}{\lambda + \mu} \qquad \ldots \ldots \quad (1.12)$$

$$k = \lambda + \frac{2\mu}{3} \qquad \ldots \ldots \quad (1.13)$$

$$\nu = \frac{\lambda}{2(\lambda + \mu)} \cdot \qquad \ldots \ldots \quad (1.14)$$

1.4 Derivation of the Wave Equation

We mentioned in section 1.2 that the equation of wavemotion is derived by considering the acceleration of a parallelepiped in the medium under the action of a net force. We assume, for this purpose, that only those stresses which act in the x direction are unbalanced (cf. Fig. 1.5). If we make the further assumption that the stresses acting on the parallelepiped are very small, then, to a first order of approximation, we may assume that the dimensions of the parallelepiped remain unchanged. Hence the net force acting on our parallelepiped in the x direction is:

$$\frac{\partial \sigma_{xx}}{\partial x} \delta x . \delta y \delta z + \frac{\partial \sigma_{xy}}{\partial y} \delta y . \delta x \delta z + \frac{\partial \sigma_{xz}}{\partial z} \delta z . \delta x \delta y$$

If we denote the density of the medium when it is unstressed by ρ_0, the mass of the parallelepiped is $\rho_0 \delta x \delta y \delta z$. Thus by Newton's second law we have, after dividing through by the volume that:

$$\frac{\partial \sigma_{xx}}{\partial x} + \frac{\partial \sigma_{xy}}{\partial y} + \frac{\partial \sigma_{xz}}{\partial z} = \rho_0 \frac{\partial^2 \xi_x}{\partial t^2} \qquad \ldots \quad (1.15)$$

If we now substitute equations (1.6) and (1.7) into equations (1.11) and then substitute these into equation (1.15) we obtain equation (1.16):

$$(\lambda + 2\mu) \frac{\partial^2 \xi_x}{\partial x^2} + \mu \frac{\partial^2 \xi_x}{\partial y^2} + \mu \frac{\partial^2 \xi_x}{\partial z^2} + (\lambda + 2\mu) \frac{\partial^2 \xi_y}{\partial x \partial y} + (\lambda + 2\mu) \frac{\partial^2 \xi_z}{\partial x \partial z}$$
$$= \rho_0 \frac{\partial^2 \xi_x}{\partial t^2} \quad (1.16)$$

As we have postulated that the displacement of the medium is in the x direction only, both ξ_y and ξ_z must be zero so that equation (1.16) reduces to:

$$(\lambda + 2\mu)\frac{\partial^2 \xi_x}{\partial x^2} + \mu\frac{\partial^2 \xi_x}{\partial y^2} + \mu\frac{\partial^2 \xi_x}{\partial z^2} = \rho_0\frac{\partial^2 \xi_x}{\partial t^2} \qquad . \qquad \textbf{(1.17)}$$

Equation (1.17) implies that two types of wavemotion can exist even though the medium displacement is in one direction only. In the first

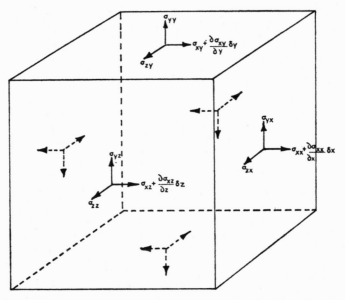

Fig. 1.5 Illustration of unbalanced stresses acting in the x direction.

case the displacement ξ_x is function only of x and t. Such waves for which the displacement is parallel to the direction of propagation are termed longitudinal waves and are described by the following equation:

$$\frac{\partial^2 \xi_x}{\partial t^2} = \left(\frac{\lambda + 2\mu}{\rho_0}\right)\frac{\partial^2 \xi_x}{\partial x^2} \qquad . \qquad . \qquad . \qquad \textbf{(1.18)}$$

The other waves in which the displacement is a function of either y or z and t travel in either the y or z directions respectively and have displacements at right angles to the direction of propagation. Such waves are termed transverse or shear waves and are described by the following equation:

$$\frac{\partial^2 \xi_x}{\partial t^2} = \left(\frac{\mu}{\rho_0}\right)\frac{\partial^2 \xi_x}{\partial y^2} \qquad . \qquad . \qquad . \qquad . \qquad \textbf{(1.19)}$$

Both equations (1.18) and (1.19) are wave equations and if we compare these with our original wave equation, equation (1.4), we see that longitudinal waves travel with a velocity c_l given by:

$$c_l = \sqrt{\frac{\lambda + 2\mu}{\rho_0}} \quad . \quad . \quad . \quad . \quad \textbf{(1.20)}$$

while transverse or shear waves have a velocity c_t given by:

$$c_t = \sqrt{\frac{\mu}{\rho_0}} \quad . \quad . \quad . \quad . \quad . \quad \textbf{(1.21)}$$

Longitudinal waves can exist in solids, liquids and gases and are the usual type of sound wave. Transverse waves are usually, but not exclusively, restricted to solids since gases and liquids do not normally possess a shear modulus except in the case of some very viscous liquids. As well as permitting both longitudinal and transverse waves solids can transmit a further type, a torsion wave, in which for example, a rotation is transmitted along a cylinder with the same velocity as that of a transverse wave. The equations for the torsion wave can be found by combining equations (1.8) with equations (1.6) and (1.7), etc.

Instead of expressing the velocities in terms of the Lamé constants it is often more convenient to employ the elastic moduli and Poisson's ratio. Thus for a solid we find:

$$c_l = \sqrt{\frac{E(1-\nu)}{\rho_0(1+\nu)(1-2\nu)}} \quad . \quad . \quad . \quad \textbf{(1.22)}$$

$$c_t = \sqrt{\frac{E}{2\rho_0(1+\nu)}} \quad . \quad . \quad . \quad . \quad . \quad \textbf{(1.23)}$$

For a liquid the appropriate modulus is the bulk modulus while the shear modulus is taken to be zero. Thus we have:

$$c_l = \sqrt{\frac{k}{\rho_0}} \quad . \quad . \quad . \quad . \quad . \quad \textbf{(1.24)}$$

$$c_t = 0 \quad . \quad . \quad . \quad . \quad . \quad \textbf{(1.25)}$$

Instead of using the bulk modulus it is sometimes convenient to use its reciprocal, the compressibility. In particular we shall be using the compressibility in chapter 6 when discussing the absorption and dispersion of sound waves.

1.5 Physical Nature of Wavemotion

At this point in our development of the mathematical theory of wavemotion it is convenient to pause in order to consider the physical nature of wavemotion as applied to acoustic waves. If we consider longitudinal waves we have a displacement of the medium, parallel to the direction in which the wave is travelling, being propagated through the medium.

The particles composing the medium do not themselves suffer any permanent displacement since they merely move backwards and forwards about their mean positions. Since the displacement is a function of x, the displacements of two planes, with their normals parallel to x, situated a short distance δx apart will not necessarily be the same at any one given time so that the passage of the wave will be accompanied by volume changes which give rise to density and pressure changes. Thus we may regard a wave as being a displacement, volume, density or pressure wave. If we were able to take, as it were, a snapshot of say the pressures existing at a given time along the x axis of a medium in which a wave is travelling and were to plot these against x we would obtain a curve which might look something like Fig. 1.6 for a periodic wave. At

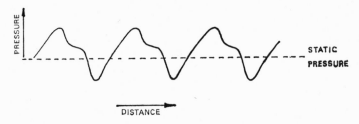

Fig. 1.6 Example of a typical periodic wavemotion.

a time t later this curve would be shifted a distance ct to the right. The curve in Fig. 1.6 is usually referred to as a waveform. We notice that the pressure fluctuates about the static pressure in the medium.

The pressure fluctuations associated with an acoustic wave usually occur fairly fast so that the heat produced during a compression does not have time to leak away before a further pressure change occurs, that is the pressure changes are adiabatic so that in the expressions for velocity we must use adiabatic values for the moduli. For example, the adiabatic bulk modulus for a gas at a static pressure P_o is γP_o where γ denotes the ratio of its specific heats so that the velocity of longitudinal waves in a gas is given by:

$$c_l = \sqrt{\frac{\gamma P_o}{\rho_o}} . \quad . \quad . \quad . \quad . \quad . \quad \textbf{(1.26)}$$

1.6 Sinusoidual Wavemotion

So far we have placed no restrictions, apart from their being very small, upon the way in which either pressure or displacement vary with time or distance. However, in ultrasonics we are concerned almost exclusively with sinusoidal motion and in the majority of applications with longitudinal wavemotion.

The complete solution of the wave equation (equation 1.4) for longitudinal sinusoidal wavemotion is:

$$\xi_x = \xi_1 \sin(\omega t - \beta x + \varphi) + \xi_2 \sin(\omega t + \beta x) \qquad \textbf{(1.27)}$$

where $\beta = \dfrac{\omega}{c_l} = \dfrac{2\pi}{\lambda}$, ω denotes the angular frequency ($2\pi f$), f the frequency, λ the wavelength (not to be confused with Lamé's first constant), ξ_1 the amplitude of a wave travelling in the positive x direction, ξ_2 the amplitude of a wave travelling in the opposite direction and φ the phase angle between the two waves. Most problems in wave theory are concerned with finding the values of ξ_1, ξ_2 and φ for given boundary conditions.

Most readers of this book will be familiar with the use of complex notation for the solution of problems in a.c. electrical networks[1.4]. This powerful technique can also be applied to ultrasonic wavemotion with the added advantage that its use enables us to find analogies between wave propagation and electrical circuits so that well-known theorems in electrical engineering can be applied directly to wavemotion. Furthermore since most ultrasonic generators and many detectors are electrical the use of complex notation technique enables us to represent both the generator or detector and the medium in which the wave is being propagated by one electrical circuit thereby easing the analysis of the complete system. Expressing equation (1.27) in complex form we have:

$$\xi_x = \xi_1 e^{j(\omega t - \beta x)} + \xi_2 e^{j(\omega t + \beta x)} \qquad . \qquad . \qquad . \qquad \textbf{(1.28)}$$

The phase angle φ has been taken into ξ_1 which will now be complex (i.e. of the form $|\xi_1|\, e^{j\varphi}$).

We can obtain expressions for the acoustic pressure p_x and the particle velocity u_x similar to that for ξ_x. If we combine the appropriate expressions for a longitudinal wave from equations (1.6) and (1.11) and note that our convention for σ_{xx} is that a tension is positive we find that:

$$p_x = -\sigma_{xx} = -(\lambda + 2\mu)\frac{\partial \xi_x}{\partial u}$$

Substituting for $\dfrac{\partial \xi_x}{\partial x}$ from equation (1.28) we obtain eventually:

$$p_x = j\beta(\lambda + 2\mu)[\xi_1 e^{j(\omega t - \beta x)} - \xi_2 e^{j(\omega t + \beta x)}] \qquad . \qquad \textbf{(1.29)}$$

From equation (1.20) we have that $\beta(\lambda + 2\mu) = \omega\rho_0 c$ where c denotes the wave velocity as there is now no need to distinguish between longitudinal and transverse waves. Thus we may now write equation (1.29) as:

$$p_x = j\omega\rho_0 c[\xi_1 e^{j(\omega t - \beta x)} - \xi_2 e^{j(\omega t + \beta x)}] \qquad . \qquad . \qquad \textbf{(1.30)}$$

Alternatively we may write equation (1.30) as:

$$p_x = P_1 e^{j(\omega t - \beta x)} + P_2 e^{j(\omega t + \beta x)} \quad . \quad . \quad . \quad \textbf{(1.31)}$$

in which $P_1 = j\omega\rho_0 c\xi_1$ and $P_2 = -j\omega\rho_0 c\xi_2$. Notice that because of the presence of j in the expressions for P_1 and P_2 the pressure leads the displacement by 90°.

In similar manner we may obtain an expression for the particle velocity $u_x = \dfrac{\partial \xi_x}{\partial t}$. We find:

$$u_x = j\omega[\xi_1 e^{j(\omega t - \beta x)} + \xi_2 e^{j(\omega t + \beta x)}] \quad . \quad . \quad \textbf{(1.32)}$$

or:

$$u_x = \frac{1}{\rho_0 c}[P_1 e^{j(\omega t - \beta x)} - P_2 e^{j(\omega t + \beta x)}] \quad . \quad . \quad \textbf{(1.33)}$$

We notice that while the particle velocity leads the displacement by 90°, pressure and particle velocity are in phase for each wave.

1.7 Energy of Sound Wave

When a sound wave is present the medium particles acquire an oscillatory motion, that is they acquire kinetic energy, so that energy is associated with a sound wave. We now require to find an expression for this energy.

We assume that the sound wave is generated by a sinusoidally oscillating piston, of area a, situated at $x = 0$. Only the wave travelling in the positive x direction is present. The displacement at any time t of the piston is given by:

$$\xi = \xi_1 \sin \omega t \quad . \quad . \quad . \quad . \quad \textbf{(1.34)}$$

and from equation (1.30) we have that the pressure exerted by the piston is:

$$p = \omega\rho_0 c\xi_1 \sin \omega t \quad . \quad . \quad . \quad . \quad \textbf{(1.35)}$$

Thus the work done over one complete cycle by the piston is given by:

$$\int pa(d\xi) = \rho_0 c\xi_1{}^2 a \int_0^{2\pi} \cos^2 \omega t\, d(\omega t)$$

which is:

$$\pi\omega\rho_0 c\xi_1{}^2 a$$

Since this is the energy input by the piston over one complete cycle it represents the energy associated with a volume of area a and length λ. If we define the energy density E as the energy in unit volume we have:

$$E = \pi\omega\rho_0 c\xi_1{}^2/\lambda \quad . \quad . \quad . \quad . \quad \textbf{(1.36)}$$

A more useful measure of energy is the intensity I which is defined as the amount of energy transmitted in one second (i.e. the power) normally through unit area. Thus we have:

$$I = Ec = \pi\omega\rho_0 c^2 \xi_1^2 / \lambda \qquad \dots \quad (1.37)$$

Using parts of equations (1.30) and 1.33) we may write alternative expressions for E and I as follows:

$$E = \frac{\rho_0 U^2}{2} = \frac{P^2}{2\rho_0 c^2} \cdot \qquad \dots \quad (1.38)$$

$$I = \frac{\rho_0 c U^2}{2} = \frac{P^2}{2\rho_0 c} \cdot \qquad \dots \quad (1.39)$$

where P and U denote the pressure and velocity amplitudes respectively

1.8 Units

The various expressions which we have so far derived are valid for any consistent system of units. While physicists still tend to prefer the C.G.S. system, engineers are rapidly transferring to the M.K.S. system which has many advantages. In this book we shall use the rationalised M.K.S. system with its fundamental units of the metre, kilogram, second and Coulomb. Table 1.1 lists the acoustical quantities and their units together with conversion factors to C.G.S. units.

TABLE 1.1

Quantity	M.K.S. Unit	Conversion to C.G.S. Unit
Length	metre	$= 10^2$ cm
Mass	kilogram	$= 10^3$ grm
Time	second	$= 1$ sec
Force	Newton	$= 10^5$ dyne
Work	Joule	$= 10^7$ erg
Power	Watt	$= 10^7$ erg/sec
Energy density	Joule/m^3	$= 10$ erg/cm^3
Intensity	Watt/m^2	$= 10^3$ erg/cm^2/sec
Pressure	Newton/m^2	$= 10$ dyne/cm^2
Density	Kgm/m^3	$= 10^{-3}$ gm/cm^3

There is one other unit which is often used to measure acoustic pressure amplitudes—the atmosphere—which equals $1 \cdot 013 \times 10^5$ Newton/m^2.

1.9 Acoustic Transmission Lines

We mentioned in section 1.6 that the use of complex notation for the analysis of acoustic wavemotion leads to an analogy with electric circuit theory enabling us to apply circuit theorems to acoustic problems. In order to develop this analogy we will consider the equations relating voltages and currents along an electrical transmission line.

An electrical transmission line consists either of two parallel wires or of two concentric cylinders between which exits a voltage and along which flows a current. A short length of each conductor will have an inductance and there will be a capacitance between the conductors. We may represent the line by the equivalent circuit of Fig. 1.7 in which

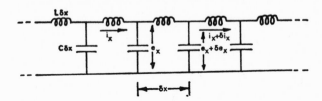

Fig. 1.7 Equivalent circuit for a section of an electrical transmission line.

each inductance has a value of $L\delta x$ and each capacitance $C\delta x$; L and C being the inductance and capacitance per unit length respectively. If we now suddenly apply a voltage to one end of the line, at $x = 0$, for example, a voltage and current wave will travel along the line. At any given t time the voltages and currents for a small section of the line will be as shown in Fig. 1.7. We may now write down the equations for the line as follows[1.5]:

$$\delta e_x = -(L\delta x)\frac{\partial i_x}{\partial t}$$

i.e.
$$\frac{\partial e_x}{\partial x} = -L\frac{\partial i_x}{\partial t} \quad . \quad . \quad . \quad . \quad . \quad (1.40)$$

and:
$$\delta i_x = -(C\delta x)\frac{\partial e_x}{\partial t}$$

i.e.
$$\frac{\partial i_x}{\partial x} = -C\frac{\partial e_x}{\partial t} \quad . \quad . \quad . \quad . \quad . \quad (1.41)$$

If we now take the following acoustic equations:

$$p_x = -(\lambda + 2\mu)\frac{\partial \xi_x}{\partial x} = -\rho_0 c^2 \frac{\partial \xi_x}{\partial x}$$

$$u_x = \frac{\partial \xi_x}{\partial t}$$

$$\frac{\partial^2 \xi_x}{\partial t^2} = c^2 \frac{\partial^2 \xi_x}{\partial x^2}$$

and combine them we eventually obtain equations (1.42) and (1.43):

$$\frac{\partial p_x}{\partial x} = - \rho_0 \frac{\partial u_x}{\partial t} \quad \cdots \quad \cdots \quad \textbf{(1.42)}$$

$$\frac{\partial u_x}{\partial x} = - \frac{1}{\rho_0 c^2} \frac{\partial p_x}{\partial t} \quad \cdots \quad \cdots \quad \textbf{(1.43)}$$

If we compare equation (1.40) with equation (1.42) and (1.41) with (1.43) we see that for each pair the acoustical and electrical equations are formally identical so that we may regard pressure, for example, as being analogous to voltage. A list of the analogies arising from equations (1.40) to (1.43) is given in table 1.2.

TABLE 1.2

Acoustical Quantity		Analogous Electrical Quantity	
pressure	p	voltage	e
particle velocity	u	current	i
displacement	ξ	charge	q
density	ρ_0	inductance/unit length	L
	$\dfrac{1}{\rho_0 c^2}$	capacitance/unit length	C

Because of this formal identity between the equations for acoustical wavemotion and electrical transmission lines we can use the well-developed theory of transmission lines directly for the solution of problems in acoustics. However, as not all readers may be familiar with transmission line theory we will derive the necessary parts of the theory from an acoustical viewpoint.

1.10 Acoustic Impedance

In electrical circuit theory the concept of impedance, e/i, plays a vital role. Similarly, because of the analogy between acoustical and electrical quantities we would expect that an analogous quantity, p/u, the acoustic impedance, will play a similar role in acoustical theory.

If we divide equation (1.31) by equation (1.33) we obtain:

$$\frac{p_x}{u_x} = Z = \rho_0 c \left[\frac{P_1 e^{-j\beta x} + P_2 e^{j\beta x}}{P_1 e^{-j\beta x} - P_2 e^{j\beta x}} \right] \quad \cdots \quad \textbf{(1.44)}$$

If we consider a system in which only the wave travelling in the positive x direction is present we find:

$$\frac{p_x}{u_x} = Z_{sp} = \rho_0 c \quad \cdots \quad \cdots \quad \textbf{(1.45)}$$

This quantity Z_{sp} is termed the specific acoustic impedance for the medium and is analogous to a pure resistance since the absence of terms involving j implies that pressure and particle velocity are in phase. This is not the case when we have both a forward- and backward-travelling wave present as equation (1.44) shows.

We have previously derived expressions for the intensity of an acoustic wave from acoustic theory but these could have been obtained directly from our acoustical–electrical analogies. If we have a resistance r through which passes a sinusoidal current of amplitude $\hat{i}$ and across which is a voltage of amplitude $\hat{e}$ the power dissipated in r is $r\hat{i}^2/2$ or $\hat{e}^2/2r$. These expressions are formally identical with equations (1.39) if we regard $\rho_0 c$ as being analogous to r. We must, however, add one caveat as our expressions for the electrical power give the total power being dissipated whereas the intensity is the power flowing through unit area. The total acoustic power is Ia where a denotes the total cross-sectional area of the wave assuming that the intensity does not vary over a. Thus we may write the total power as $(Pa)^2/2\rho_0 ca$. This indicates that we should regard $\rho_0 c$ as the impedance per unit area so that the total impedance is $\rho_0 ca$ in which case the acoustical analogy of voltage is Pa, that is, force. Thus the analogies given in table 1.2 are really those applicable to a wave of unit cross-sectional area.

1.11 Standing Waves in Transmission Lines

So far we have considered only the presence of both forward- and backward-travelling waves but not how these are produced. In general we have a generator radiating a plane wave, that is one in which the intensity is constant across the front of the wave except at the outer edges where it falls to zero, which, after travelling a distance l is reflected back on itself by some reflecting surface. This reflected wave returns to the generator where it is again reflected so that it again is travelling towards the reflector where it is again reflected and so on. Thus in the space between the generator and reflector we have a set of forward- and backward-travelling waves. At equilibrium we can sum each set of waves into one wave so that we effectively have just one forward- and one backward-travelling wave as was expressed mathematically by equation (1.27). The pressure at any point x can be found from equation (1.31). For example, if we assume $P_1 = P_2$ equation (1.31) leads to:

$$p_x = 2P_1 \cos \beta x \,.\, e^{j\omega t} \quad .\quad .\quad .\quad .\quad \textbf{(1.46)}$$

Equation (1.46) shows that the pressure at any point x oscillates sinusoidally with an amplitude of $2P_1 \cos \beta x$, unlike the situation which occurs with only a forward-travelling wave where the pressure at one point oscillates at the same frequency but with an amplitude independent of x. A pressure distribution given by equation (1.46) and illustrated in Fig. 1.8 is known as a standing wave since there is no net transport of energy $(P_1 = P_2)$*.

1.12 Input Impedance of a Transmission Line

Equation (1.44) gives the impedance at any point x in the presence of both forward and backward waves. We have shown in the previous section that the backward-travelling wave comes about by reflection. This situation arises in practice when, for example, a sound wave beam

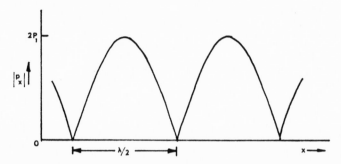

Fig. 1.8 Pressure distribution in a standing wave formed by complete reflection of the incident wave.

passes through a plate and the analysis of such situations is facilitated by knowing the impedance at the input side of the plate.

We consider a system in which we have a generator of acoustic waves situated at one end of an acoustic transmission line which is terminated a distance l away from the generator by a medium of specific acoustic impedance Z_t. The specific acoustic impedance of the medium forming the line is denoted by Z_o. Thus the impedance at a point x is given by equation (1.44) to be:

$$Z_x = Z_o \left[\frac{P_1 e^{-j\beta x} + P_2 e^{j\beta x}}{P_1 e^{-j\beta x} - P_2 e^{j\beta x}} \right] \quad \cdot \quad \cdot \quad \cdot \quad (1.47)$$

At the termination, for which $x = l$, we have that $Z_x = Z_t$ whence from equation (1.47) we have:

$$P_2 = P_1 e^{-j2\beta l} \left[\frac{Zt - Z_o}{Zt + Z_o} \right] \quad \cdot \quad \cdot \quad \cdot \quad (1.48)$$

If we now substitute equation (1.48) into equation (1.47) we can eliminate P_1 and P_2 to obtain:

$$Z_x = Z_o \frac{Zt + jZ_o \tan \beta(l - x)}{Z_o + jZ_t \tan \beta(l - x)} \quad \cdot \quad \cdot \quad (1.49)$$

* Standing waves can still ensue even if $P_1 \neq P_2$ and there is a net energy transport.

Thus the input impedance Z_{in} of the line is:

$$Z_{in} = Z_o \frac{Z_t + jZ_o \tan \beta l}{Z_o + jZ_t \tan \beta l} \quad \cdot \quad \cdot \quad \cdot \quad \textbf{(1.50)}$$

The impedance expressions derived above are valid only for the situations in which there is no loss of acoustic energy by absorption in the medium in which the acoustic waves are travelling. Should this absorption not be negligible the term $j\beta$ must be replaced by $(\alpha + j\beta)$ where α denotes the absorption coefficient for the medium. In this case we have:

$$Z_x = Z_o \frac{Z_t + Z_o \tanh[(\alpha + j\beta)(l - x)]}{Z_o + Z_t \tanh[(\alpha + j\beta)(l - x)]} \quad \cdot \quad \cdot \quad \textbf{(1.51)}$$

1.13 Impedance Transformers

There are two special cases of equation (1.50) which are important: the first in which the line is a quarter-wavelength long and the second in which it is a half-wavelength long.

Quarter Wave Transformer

If the line is a quarter-wavelength long $\tan \beta l = \infty$ so that equation (1.50) gives:

$$Z_{in} = Z_o{}^2/Z_t \quad \cdot \quad \cdot \quad \cdot \quad \cdot \quad \textbf{(1.52)}$$

Half Wave Transformer

If the line is a half-wavelength long $\tan \beta l = 0$ so that equation (1.50) gives:

$$Z_{in} = Z_t \quad \cdot \quad \cdot \quad \cdot \quad \cdot \quad \cdot \quad \textbf{(1.53)}$$

These impedance transformers are important since it is sometimes necessary to convert a high impedance to a low value, or vice versa, in which cases one would use a quarter-wave line made of some suitable medium while in other cases it is necessary that a given impedance be coupled by a length of line to say the end of another line in such a way that the impedance appearing at the end of this line is the same as the original given impedance. In this case one would use a half-wave transformer. These points are dealt with in greater detail when we consider transmission of acoustic waves through plates.

1.14 Reflection and Refraction at a Boundary

We must now consider the problem of what happens when an acoustic wave travelling in one medium with a given specific acoustic impedance meets a boundary on the other side of which is a medium with a differing impedance. In general part of the incident energy is reflected and part refracted with the ratio being governed by the media impedances and the angle of incidence of the waves on to the boundary. As we are not restricting the angle of incidence to zero (i.e. normal incidence) we cannot use transmission line theory; instead we have to consider the boundary conditions at the interface between the two media. These boundary conditions are:

1. Continuity of normal and tangential stresses.
2. Continuity of normal and tangential velocities and displacements.

If the first condition were not satisfied a stress would act on the boundary and would generate a wave. However, this wave already exists as the refracted wave and hence the stress must be continuous. The second condition merely states that a particle situated in the boundary plane can have only one velocity and displacement at any given time.

The angles between the directions of propagation of the incident, reflected and refracted waves and the boundary normal are governed, as in optics, by Snell's law. To show this we consider the system illustrated in Fig. 1.9. If we consider, say, one compression peak of the incident wave this will meet different parts of the boundary at different

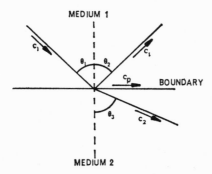

Fig. 1.9 Incomplete representation of reflection and refraction at a boundary.

times—in fact this peak will appear to travel along the boundary with a velocity $c_1/\sin \theta_1$ where c_1 is the acoustic velocity in medium 1. This velocity is the velocity of propagation of a point of constant phase along the boundary and as such is termed the phase velocity c_p. As the incident wave generates the reflected and refracted waves we may reasonably expect that these three waves will all have the same phase velocity along the boundary. These reflected and refracted waves can be both longitudinal and transverse for a purely longitudinal incident wave. Thus if we denote quantities appertaining to transverse waves by primed symbols we have, for a purely longitudinal incident wave, that:

$$c_p = \frac{c_1}{\sin \theta_1} = \frac{c_1}{\sin \theta_2} = \frac{c_2}{\sin \theta_3} = \frac{c_1'}{\sin \theta_2'} = \frac{c_2'}{\sin \theta_3'} \quad (1.54)$$

We consider now the relationships between the three waves of Fig. 1.9 as set by our boundary conditions. Thus denoting stress by $-\sigma$ and noting that the positive particle velocity for a wave travelling in medium 1 in the positive x direction is $- \sigma/Z_1$, while that for a wave travelling in the opposite direction is σ/Z_1, we have for the boundary conditions:

Normal stresses $\qquad - \sigma_1 \cos \theta_1 - \sigma_2 \cos \theta_2 = - \sigma_3 \cos \theta_2$ **(1.55a)**

Tangential stresses $\qquad - \sigma_1 \sin \theta_1 + \sigma_2 \sin \theta_2 = - \sigma_3 \sin \theta_3$ **(1.55b)**

Normal velocities $\qquad - \dfrac{\sigma_1}{Z_1} \cos \theta_1 + \dfrac{\sigma_2}{Z_1} \cos \theta_2 = - \dfrac{\sigma_3}{Z_2} \cos \theta_3$ **(1.55c)**

Tangential velocities $- \dfrac{\sigma_1}{Z_1} \sin \theta_1 - \dfrac{\sigma_2}{Z_1} \sin \theta_2 = - \dfrac{\sigma_3}{Z_2} \sin \theta_3$ **(1.55d)**

As the relationships between the angles θ_1, θ_2 and θ_3 are known from equations (1.54) we have four simultaneous independent equations but only two unknowns, σ_2/σ_1 and σ_3/σ_1. However, such a set of equations requires four unknown quantities for a complete solution. Alternatively the solutions for σ_2/σ_1 and σ_3/σ_1 from any one pair of equations do not agree with those obtained from the remaining pair. We conclude, therefore, that in order to obtain solutions for equations (1.55a) to (1.55d) two more terms must be inserted into each equation and as each term must represent a wave the picture given in Fig. 1.9 must be incomplete. Also as these additional waves must travel at different angles to those in Fig. 1.9 they must be transverse waves. Thus the complete representation of the position at a boundary between two isotropic media will be as shown in Fig. 1.10.

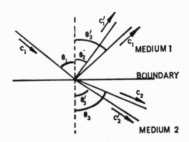

Fig. 1.10 Complete representation of reflection and refraction at a boundary.

The complete solution for reflection and refraction at a boundary is complicated since the incident wave may be either longitudinal or transverse and in the latter case the plane of polarisation can be either parallel or normal to the boundary. Also as the angle of incidence is varied the amplitudes of the reflected and refracted waves also vary. Thus here we can only point out the salient features of what happens at a boundary; for a much fuller treatment readers should consult Ergin[1.8] who has calculated the amplitudes of the reflected and refracted waves at a fluid–solid boundary (this has similar properties to a solid–solid boundary) and one set of his curves is reproduced in Fig. 1.11. The important point which these curves demonstrate is that for each refracted wave there is a critical angle for the incident wave beyond which the refracted wave does not exist. This critical angle is given by equations (1.54) when

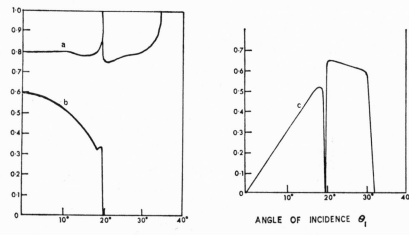

ANGLE OF INCIDENCE θ_1

Fig. 1.11 Partition of energy on reflection and refraction of a longitudinal wave at a fluid–solid boundary.
(*a*) Reflection coefficient for the reflected longitudinal wave.
(*b*) Transmission coefficient for the refracted longitudinal wave.
(*c*) Transmission coefficient for the refracted transverse wave. (After Ergin[1.8]).

the refracted wave's angle reaches 90°. Thus the critical angle of incidence for a longitudinal wave beyond which the transverse refracted wave does not exist is given by:

$$\sin \theta_1 = c_2'/c_1 \quad . \quad . \quad . \quad . \quad (1.56)$$

For angles of incidence greater than critical for both the longitudinal and transverse refracted waves, the incident longitudinal wave is reflected without loss but with a phase change.

1.15 Reflection and Refraction at Normal Incidence

One case of the general theory outlined in the previous section which is easily soluble is that of a longitudinal wave normally incident on the boundary. In this situation no transverse waves can be generated by the incident wave since there is no stress component parallel to the boundary (we neglect coupling due to Poisson's ratio) so that the four boundary equations (1.55) reduce to two, viz.:

$$- \sigma_1 - \sigma_2 = - \sigma_3 \quad . \quad . \quad . \quad . \quad (1.57a)$$

$$- \frac{\sigma_1}{Z_1} + \frac{\sigma_2}{Z_1} = - \frac{\sigma_3}{Z_2} \quad . \quad . \quad . \quad . \quad (1.57b)$$

Solving these equations leads to:

$$\frac{\sigma_2}{\sigma_1} = \frac{Z_2 - Z_1}{Z_2 + Z_1} = \frac{m - 1}{m + 1} \quad \cdots \quad \textbf{(1.58)}$$

$$\frac{\sigma_3}{\sigma_1} = \frac{2Z_2}{Z_2 + Z_1} = \frac{2m}{m + 1} \quad \cdots \quad \textbf{(1.59)}$$

where $m = Z_2/Z_1$.

The corresponding expressions for intensity reflectivity and transmissivity are:

$$\frac{I_2}{I_1} = \left(\frac{m - 1}{m + 1}\right)^2 \quad \cdots \quad \textbf{(1.60)}$$

$$\frac{I_3}{I_1} = \frac{4m}{(m + 1)^2} \quad \cdots \quad \textbf{(1.61)}$$

1.16 Transmission through Plates (Normal Incidence)

It is sometimes necessary, for mechanical reasons, to place a parallel-sided plate in the path of a beam of acoustic waves travelling in a liquid. Under this circumstance we require to know the energy loss occurring due to reflections at the liquid—solid interfaces. In order to calculate the energy transmission we consider the plate to be a lossless transmission line so that, by equation (1.50), its input impedance is:

$$Z_{in} = mZ_1 \frac{1 + jm \tan \beta l}{m + j \tan \beta l} \quad \cdots \quad \textbf{(1.62)}$$

where $m = Z_0/Z_1$, Z_0 being the specific acoustic impedance of the plate medium and Z_1 that of the liquid on either side of the plate and l is the plate thickness. We now use equation (1.59), with $Z_{in} = Z_2$, to find the pressure reflectivity. As this will usually be complex we square its modulus in order to find the intensity reflectivity R. After finding R we can easily find the intensity transmissivity T since $T = 1 - R$. Thus we find:

$$T = \frac{1}{1 + \frac{1}{4}\left(m - \dfrac{1}{m}\right)^2 \sin^2 \beta l} \quad \cdots \quad \textbf{(1.63)}$$

$$R = \frac{\frac{1}{4}\left(m - \dfrac{1}{m}\right)^2 \sin^2 \beta l}{1 + \frac{1}{4}\left(m - \dfrac{1}{m}\right)^2 \sin^2 \beta l} \quad \cdots \quad \textbf{(1.64)}$$

The variation of T with l is shown in Fig. 1.12 which shows that for certain values of l, given by $\sin \beta l = 0$ (i.e. $l = \dfrac{n\lambda}{2}$, $n = 0, 1, 2, 3 \ldots$)

the transmissivity theoretically equals unity. In effect the plate is now acting as a half-wave transformer so that its input impedance equals that of the liquid so that no reflection occurs.

1.17 Doppler Effect

The Doppler effect is the name given to the fact that when either the acoustic generator or an observer moves relative to the medium in which the waves travel, the observer measures a different acoustic frequency to that when no motion occurs.

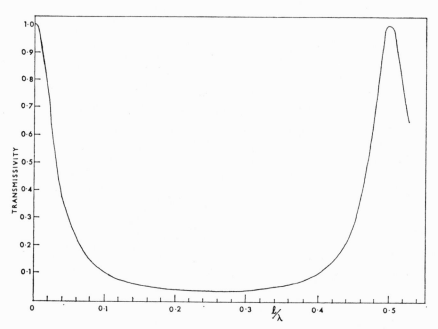

Fig. 1.12 Transmissivity of a plate of thickness l as a function of l/λ.

We consider first the case of the observer moving relative to the medium, with a velocity v_0, in the opposite direction to that in which the waves are moving. We will use the subscript 1 to denote quantities measured by a stationary observer and the subscript 2 for those measured by the moving observer. Thus for a plane wave travelling in the positive x direction we have:

$$\xi_1 = \hat{\xi}_1 \sin(\omega_1 t - \beta_1 x_1) \quad . \quad . \quad . \quad . \quad \textbf{(1.65)}$$

$$\xi_2 = \hat{\xi}_2 \sin(\omega_2 t - \beta_2 x_2) \quad . \quad . \quad . \quad . \quad \textbf{(1.66)}$$

Also:

$$x_2 = x_1 + v_0 t \quad . \quad . \quad . \quad . \quad . \quad \textbf{(1.67)}$$

Equation (1.67) shows that both observers agree on the distance between two given points along the wave, that is $\lambda_1 = \lambda_2$ and $\hat{\xi}_1 = \hat{\xi}_2 = \xi_0$ say.

If we now substitute equation (1.67) into equation (1.65) we obtain eventually:

$$\xi_2 = \xi_o \sin[\omega_1\left(1 + \frac{v_0}{c_1}\right)t - \beta_2 x_2] \quad . \quad . \quad . \quad \textbf{(1.68)}$$

Comparing equations (1.66) and (1.68) we see that:

$$\omega_2 = \omega_1\left(1 + \frac{v_0}{c_1}\right) \quad . \quad . \quad . \quad . \quad . \quad \textbf{(1.69)}$$

Thus we find that an observer moving against the acoustic wave measures a higher frequency than a stationary one while one moving with the wave measures a lower frequency given by $\omega_1\left(1 - \frac{v_0}{c_1}\right)$.

We now consider the case of a generator moving with a velocity v_g in the same direction as the waves. An observer, stationary relative to the medium, will observe a frequency ω_1. An observer sitting on the generator would, from equation (1.69), observe a frequency $\omega_1\left(1 - \frac{v_g}{c_1}\right)$ which will be the frequency of the sound source which we denote by ω_g. Hence we have:

$$\omega_1 = \frac{\omega_g}{1 - v_g/c_1} \quad . \quad . \quad . \quad . \quad . \quad \textbf{(1.70)}$$

Thus we find that the frequency of the waves emitted by a generator moving towards an observer is measured by that observer to be higher than if the generator were stationary.

1.18 Group Velocity

So far in this chapter we have tacitly assumed that the ultrasonic waves exist as a continuous sine wave of constant amplitude. For many applications, however, the sine wave is amplitude modulated, that is its amplitude is caused to vary in a predetermined manner with both distance and time. This modulated wave can be regarded as being composed of a number of continuous sine waves suitably related in frequency, amplitude and phase which may be found by means of Fourier analysis of the composite waveform. If the modulation is repetitive with a fixed frequency and we denote the modulated waveform by $F(t, x)$ we have:

$$F(t, x) = \sum_n A_n e^{j\omega_n(t-x/c_n)} \quad . \quad . \quad . \quad \textbf{(1.71)}$$

where A_n, ω_n and c_n denote the complex amplitude, angular frequency and velocity for the nth Fourier component. If the medium through which the waves are travelling displays a frequency dependent absorption and ultrasonic velocity the relative amplitudes and phases of the Fourier components will change as the waveform progresses so that it will become distorted. In addition since each component travels with a velocity dependent upon its frequency the velocity of propagation of the modulation must be some sort of average of the components' velocities and is known as the group velocity. We can derive a simple expression for this group velocity if we restrict the analysis to the case in which the medium possesses zero absorption and the Fourier components all have frequencies which do not differ much from the unmodulated wave frequency $\omega_0/2\pi$. Thus we may write:

$$\omega_n = \omega_o + \omega_i \qquad \qquad \textbf{(1.72)}$$

$$c_n = c_o + \left(\frac{\partial c}{\partial \omega}\right)_{\omega_o} \omega_i \qquad \qquad \textbf{(1.73)}$$

If we insert equations (1.72) and (1.73) into equation (1.71) and, for convenience denote A_n now by A_i we find, provided that $\omega_i \ll \omega_o$, that:

$$F(t, x) = \left[\sum_n A_i e^{j\left(\omega_i t - \frac{\omega_i x}{c_o}\left(1 - \frac{\omega_o}{c_o}\frac{\partial c}{\partial \omega}\right)\right)}\right] e^{j\omega_o(t - x/c_o)} \qquad \textbf{(1.74)}$$

Equation (1.74) shows that we may represent the signal mathematically by the product of two terms one being the standard exponential representation of a sine wave and the other, enclosed within the square brackets, being the modulation envelope. This latter term shows that the modulation travels with a velocity, the group velocity c_g, given by:

$$c_g = c_o + \omega_o\left(\frac{\partial c}{\partial \omega}\right)_{\omega_o} = \frac{\partial (c_o f_o)}{\partial f_o} \qquad \qquad \textbf{(1.75)}$$

If we consider the case in which the modulation takes the form of a single pulse the energy being propagated is contained within this pulse. Thus we have that the energy associated with a modulated wave travels at the group velocity and not with the phase velocity. This effect is important when pulse-modulated waves are transmitted through solid rods since because of the coupling between different modes the phase velocity is frequency dependent.

References

1.1 Southwell, R. V., *Theory of Elasticity*, p. 315. Oxford (1941).
1.2 Kolsky, H., *Stress Waves in Solids*, p. 8. Oxford (1953).
1.3 Ibid., p. 9.

1.4 Smirnov, V. I., *A Course of Higher Mathematics*, Vol. 1, p. 470. Pergamon (1964).

1.5 Koehler, G., *Circuits and Networks*, p. 217. Macmillan (1955).

1.6 Kolsky, H., *Stress Waves in Solids*, p. 27. Oxford (1953).

1.7 Macelwane, J. B. and Sohan, F. W., *Introduction to Theoretical Seismo-* *logy*. Wiley (1936).

1.8 Ergin, K., *Bull. Seismological Soc. Amer.*, **42**, 349 (1952).

2
Radiation

In the previous chapter we obtained equations describing the behaviour of longitudinal waves under the assumption that these were plane waves, that is, that the phase and amplitude of the waves in a plane perpendicular to the direction of propagation were constant over the plane. As we shall see, this assumption implies that the generator has an infinite-radiating area which, in practice, cannot be the case. Nevertheless, plane-wave theory is often sufficiently accurate for normal use, but in some circumstances neglect of the fact that a finite-sized generator does not generate plane waves can lead to errors. In this chapter we will consider the properties of waves radiating from a finite generator.

2.1 Huyghen's Principle

Huyghen's principle is fundamental to radiation theory. It asserts that any wave phenomenon can be analysed by the addition of contributions from some distribution of simple sources properly selected in phase and amplitude to represent the physical situation. Each simple source is usually assumed to radiate uniformly over a solid angle of 2π radians in the chosen forward direction. Hence before we can apply this principle we must first consider some properties of spherical radiation.

2.2 Spherical Radiation

We consider a disturbance travelling in an arbitrary direction relative to a rectangular co-ordinate system. The position of the disturbance at any time is given by $lx + my + nz$ where l, m and n are the direction cosines of the direction in which the disturbance is travelling. Thus we may now rewrite equation (1.2) as :

$$\theta = f(lx + my + nz - ct) \quad . \quad . \quad . \quad (2.1)$$

If we now differentiate equation (2.1) twice with respect to t and then to x we find:

$$\frac{\partial^2\theta}{\partial x^2} + \frac{\partial^2\theta}{\partial y^2} + \frac{\partial^2\theta}{\partial z^2} = \frac{1}{c^2}\frac{\partial^2\theta}{\partial t^2} \quad . \quad . \quad . \quad (2.2)$$

Equation (2.2) is the wave equation for motion in three dimensions: equation (1.4) is a special case of equation (2.2).

For spherical waves it is more convenient to express equation (2.2) in spherical co-ordinates r, θ and ψ where r denotes the distance from the origin and θ and ψ denote two angles as shown in Fig. 2.1. Transforming to spherical polar co-ordinates gives the wave equation as:

$$\frac{\partial^2 p_r}{\partial r^2} + \frac{2}{r}\frac{\partial p_r}{\partial r} + \frac{1}{r^2 \sin\theta}\frac{\partial}{\partial\theta}\left(\sin\theta \cdot \frac{\partial p_r}{\partial\theta}\right) + \frac{1}{r^2 \sin^2\theta}\frac{\partial^2 p_r}{\partial\psi^2} = \frac{1}{c^2}\frac{\partial^2 p_r}{\partial t^2}$$
(2.3)

in which we have replaced the parameter θ by pressure p_r in order to avoid confusion with the angle θ.

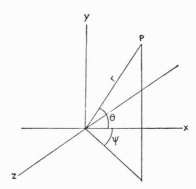

Fig. 2.1 Spherical co-ordinates.

As the simple sources mentioned in the statement of Huyghen's principle are required to radiate uniformly over the forward direction the wave equation applicable to these will be equation (2.3) with the terms implying a functional relationship between p_r and θ and ψ made zero. Thus the wave equation for spherically symmetrical radiation is:

$$\frac{\partial^2 p_r}{\partial r^2} + \frac{2}{r}\frac{\partial p_r}{\partial r} = \frac{1}{c^2}\frac{\partial^2 p_r}{\partial t^2} \qquad \cdot \qquad \cdot \qquad \cdot \qquad \textbf{(2.4)}$$

The general solution to equation (2.4) is:

$$p_r = \frac{1}{r}F(\omega t - \beta r) + \frac{1}{r}F(\omega t + \beta r) \cdot \quad \cdot \quad \cdot \quad \textbf{(2.5)}$$

If we restrict ourselves to a forward-travelling sinusoidal wave the solution to equation (2.4) becomes:

$$p_r = P_a \frac{a}{r} e^{j(\omega t - \beta r)} \quad \cdot \quad \cdot \quad \cdot \quad \cdot \quad \textbf{(2.6)}$$

where P_a denotes the pressure amplitude at a distance a from the centre of the source which may, for example, be regarded as a pulsating sphere.

Spherically symmetrical waves differ from plane waves in that (*a*) their wave front is the surface of a sphere instead of a plane. (*b*) the pressure amplitude decreases as the wave advances instead of remaining constant (in the absence of absorption) and (c) the specific acoustic impedance is complex instead of being completely real. To show this last point we can take the expressions derived in chapter 1 for pressure and particle velocity as a function of displacement and combine these with the wave equation to give:

$$u_r = \frac{j}{\omega\rho_0}\frac{\partial p_r}{\partial r} = \frac{p_r}{\rho_0 c}\left[1 + \frac{1}{j\beta r}\right] \qquad \cdots \qquad (2.7)$$

thus the specific acoustic impedance Z_r is:

$$Z_r = \frac{p_r}{u_r} = \frac{\rho_0 c}{1 + \dfrac{1}{j\beta r}} \qquad \cdots \qquad (2.8)$$

Although Z_r is complex its magnitude rapidly approaches the plane-wave impedance as r increases since the difference between the two is less than 1% for $r > 1.1\lambda$.

2.3 Radiation from a Circular Piston

Most acoustic generators take the form of a flat plate whose front surface vibrates parallel to the surface normal thereby generating longitudinal waves. The mathematical analysis of the radiation from such a generator is greatly facilitated if we make certain assumptions, which, while they are not always met in practice, nevertheless lead to useful results. The shape of the radiating surface can be either circular or

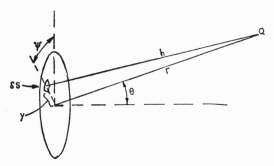

Fig. 2.2 Construction used in the derivation of the radiation pattern from a circular piston.

rectangular but we will confine our study to circular radiators as these are the commonest. We assume, therefore, that the radiator is a circular piston set in a tight fitting circular hole in an infinite plate or baffle whose function is to prevent radiation from the rear of the piston travelling round to the front and interfering with the forward-radiated energy.

We also assume that the front surface of the piston vibrates uniformly, that is that every point vibrates with the same amplitude and phase as every other. This is not the case with, for example, a quartz disc, which usually vibrates non-uniformly with a variety of amplitudes distributed over its surface.

In order to find how the radiation is distributed in front of the generator we have to find the pressure at a point Q (cf. Fig. 2.2) situated a distance r from the centre of the piston. To do this we first of all find the contribution to the pressure at Q made by a small area of the piston, δS, distant h from Q and then integrate over the total radiating surface of the piston. In order to find the contribution made by δS we consider two infinitesimally small radiating spheres placed side by side. If h is sufficiently large the pressure at Q is $\dfrac{2P_a a}{h} e^{j(\omega t - \beta h)}$, that is the contribution of two spheres to the pressure at Q is twice that of one and still varies inversely with distance h. But the area of two spheres is twice that of one sphere, therefore we can say that the contribution to the pressure at Q of a small area δS, is directly proportional to δS and inversely proportional to h. Therefore the total pressure P_Q at Q is given by:

$$P_Q = \int_s \frac{P'}{h} e^{j(\omega t - \beta h)} \, dS \; . \qquad . \quad . \quad . \quad (2.9)$$

where P' is proportional to the pressure at the face of the piston.

Also:
$$\delta S = y \delta \psi \delta y$$

and:
$$h = (r^2 + y^2 - 2ry \sin \theta \cdot \cos \psi)^{\frac{1}{2}}$$

In principle we can now solve the integral of equation (2.9) but in practice this can only be done by making certain approximations. To start with we will consider two special cases (*a*) when $r \gg r_o$ and (*b*) the pressure variation along the normal to the piston surface and which passes through the piston's centre.

2.3a The Far Field Case (Fraunhofer region)

In this case, for which $r \gg r_o$, we can write:

$$h \approx r - y \sin \theta \cos \psi$$

As a further approximation we write:

$$\frac{1}{h} \approx \frac{1}{r}$$

Thus:

$$P_Q = \frac{P'}{r} e^{j(\omega t - \beta r)} \int_0^{r0} y \, dy \int_0^{2\pi} e^{j\beta y \sin \theta \cos \psi} \, d\psi \; . \qquad (2.10)$$

Integrating gives:

$$P_Q = \frac{P'\pi r_o^2}{r} e^{j(\omega t - \beta r)} \left[\frac{2J_1(\beta r_o \sin \theta)}{\beta r_o \sin \theta} \right] \qquad . \quad . \quad \textbf{(2.11)}$$

where J_1 is the Bessel function of the first kind.

Equation (2.11) shows that the pressure at distances from the piston which are large compared with the piston radius is a complicated function of θ as well as decreasing with increasing distance. The term within the square brackets in equation (2.11) is known as the directivity function as it gives the variation in pressure with direction. This term for the circular piston is plotted in Fig. 2.3. Directivity functions for other types of generators are given in table 2.1.

TABLE 2.1. *Directivity functions for various sources. Each distributed or point source has uniform strength and phase.*

Source Geometry	Directivity Function	x
N point sources	$\dfrac{\sin (Nx)}{N \sin (x)}$	$\dfrac{\pi d}{\lambda} \sin \theta$
Distributed line	$\dfrac{\sin (x)}{x}$	$\dfrac{\pi l}{\lambda} \sin \theta$
Annulus	$J_o(x)$	$\dfrac{2\pi r_o}{\lambda} \sin \theta$
Plane circular piston	$\dfrac{2J_1(x)}{x}$	$\dfrac{2\pi r_o}{\lambda} \sin \theta$
Plane rectangular piston	$\dfrac{\sin (x_\theta) \sin (x_\varphi)}{x_\theta \quad x_\varphi}$	$x_\theta = \dfrac{\pi a}{\lambda} \sin \theta$ $x_\varphi = \dfrac{\pi b}{\lambda} \sin \theta$

An alternative way of displaying the variation in pressure with direction is the polar diagram. A polar diagram is the locus of one end of a vector, the other end of which is situated at the generator centre, whose direction is the same as that in which we wish to find the pressure and whose length is proportional to the pressure. Examples of polar diagrams for a piston generator are given in Fig. 2.4. These show that as

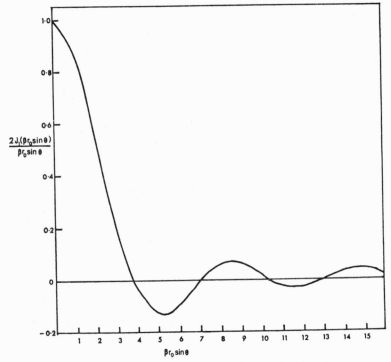

Fig. 2.3 Plot of the directivity function for a circular piston.

the ratio of diameter to wavelength increases the angular divergence of the beam or lobe of the radiation decreases but that on further increasing $\frac{2r_o}{\lambda}$ small side lobes appear. If we regard the main lobe as being contained within a cone of semiangle θ_c this angle is given by the smallest value of $\beta r_o \sin \theta$ which makes the function plotted in Fig. 2.3 become zero. Thus we have:

$$\frac{2\pi r_o}{\lambda} \sin \theta_c = 3.832$$

that is:

$$\sin \theta_c \approx 0.61 \frac{\lambda}{r_o} \quad \cdot \quad \cdot \quad \cdot \quad \cdot \quad (2.12)$$

While these side lobes can be important in, for example, flaw detection where it is important that all the radiation is confined within as narrow a beam as possible pointing in a specified direction, they contain less than 20% of the radiated energy.

2.3b Axial pressure variation in front of circular piston

If we restrict the position of Q in Fig. 2.2 to lie on the normal passing through the piston centre equation (2.10) now becomes:

$$P_Q = P' e^{j\omega t} \int_0^{2\pi} \partial\psi \int_0^{r_0} \frac{e^{j\beta\sqrt{x^2+y^2}}}{\sqrt{x^2 + y^2}} y \, dy \quad . \quad . \quad (2.13)$$

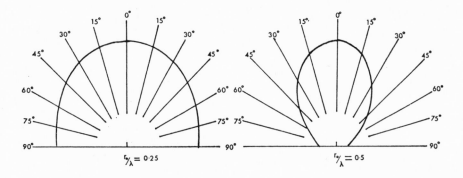

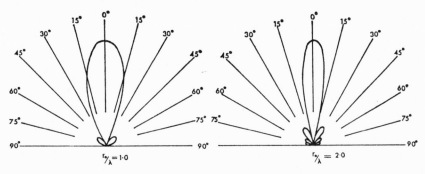

Fig. 2.4 Polar diagrams for the radiation from circular pistons. (After Hueter and Bolt, Sonics, 65.)

After integrating and some manipulation we find:

$$|P_Q| = \frac{2\pi P'}{\beta} [2 - 2 \cos \beta(\sqrt{r^2 + x^2} - x)]^{\frac{1}{2}} \quad . \quad (2.14)$$

The form taken by the magnitude of P_Q, $|P_Q|$, is shown in Fig. 2.5. The region in which P_Q fluctuates rapidly is known as the near field or Fresnel region in which the condition for either a maximum or minimum in $|P_Q|$ is $\cos \beta(\sqrt{r^2 + x^2} - x) = -1$ or $+1$ respectively, that is:

$$x = \frac{4r_0^2 - n^2\lambda^2}{4n\lambda} \quad . \quad . \quad . \quad . \quad (2.15)$$

where $n = 1, 3, 5, 7$ . . . . . . for a maximum
             $n = 2, 4, 6, 8$ . . . . . . for a minimum

For distances greater than that at which the last maximum occurs (i.e. when $n = 1$), P_Q decreases monotonically as x increases, that is we pass into the Fraunhofer region. If we make $x \gg r_0$ in equation (2.14)

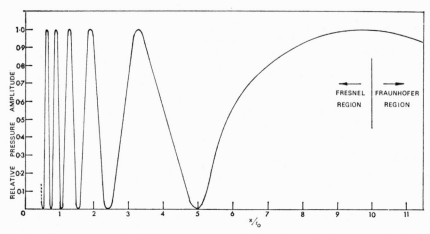

Fig. 2.5 The pressure distribution along the axis of radiation for a circular piston of radius r_0.

and $\theta = 0$ in equation (2.11) we obtain the same expression for the axial pressure variation at large distances from the source, viz.:

$$| P_Q | = \frac{\pi r_0^2 P'}{x} \quad . \quad . \quad . \quad . \quad (2.16)$$

2.3c General solution for radiating piston

Although we have restricted the discussion so far to two special cases because they are amenable to calculation, a general solution is obviously desirable. Several workers [2.1], [2.2], [2.3] have given solutions to equation (2.9) in the form of approximation formulae suitable for numerical computation. As one example we consider the expression given by Seki, Granato and Truell[2.3] for the pressure at Q. They show that if one takes the pressure which would exist in the Fraunhofer region, were the piston to radiate a perfect plane wave as unity, the pressure P_Q' at Q is given by:

$$P_Q' = (X^2 + Y^2)^{\frac{1}{2}} \cos (\beta ct - \beta x - \gamma(x, z)) \quad . \quad (2.17)$$

where

$$X = V_0 \cos \left[\frac{\beta(z^2 + r_0^2)}{2x} \right] + V_1 \sin \left[\frac{\beta(z^2 + r_0^2)}{2x} \right] - 1$$

$$Y = V_0 \sin\left[\frac{\beta(z^2 + r_0^2)}{2x}\right] - V_1 \cos\left[\frac{\beta(z^2 + r_0^2)}{2x}\right]$$

$$V_0 = \sum_{n=1}^{\infty}(-1)^n \left(\frac{z}{r_0}\right)^{-2n} J_{2n}\left(\frac{\beta r_0 z}{x}\right)$$

$$V_1 = \sum_{n=0}^{\infty}(-1)^n \left(\frac{z}{r_0}\right)^{-2n-1} J_{2n+1}\left(\frac{\beta r_0 z}{x}\right)$$

$$\gamma(x, z) = \tan^{-1}\left(\frac{Y}{X}\right)$$

A typical set of plots of the pressure amplitude $(X^2 + Y^2)^{\frac{1}{2}}$ and phase $\gamma(x, z)$ is given in Fig. 2.6. One point which these curves demonstrate

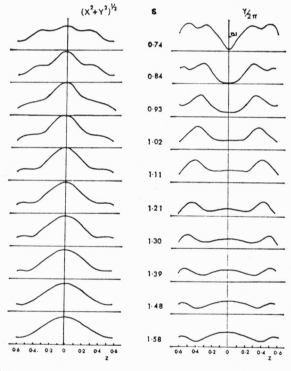

Fig. 2.6 Maximum pressure and phase-angle variations across planes normal to the direction of propagation at various distances from a circular piston. s is expressed in units of r^2/λ. (After Seki *et al.* (2.3).)

is that in the Fresnel region the radiated wave is not plane, as is often assumed, since both the amplitude and phase vary across the wavefront. This variation in amplitude across the wavefront can be important in, for example, measurements of the absorption of ultrasonic waves since the normal technique involves measuring the variation in pressure amplitude with increasing distance from the generator. The pressure measuring device usually takes the form of a flat disc which measures the average pressure across its surface area. Seki, Granato and Truell have given the average r.m.s. pressure over a circular plane surface identical to that of the generator for values of βr_o and distance S expressed in terms of r_o^2/λ. Their results are reproduced in table 2.2. which also includes values

TABLE 2.2. *Average r.m.s. pressure over surface of detecting piston as a function of distance* S (S *in units of* r_o^2/λ)

S	$\beta r_o = 50$	$\beta r_o = 100$	$\beta r_o = 200$	$\beta r_o = 1000$	db loss
0·2	1·04198	0·8860	0·9053	0·9057	0·868
0·3	0·8612	0·8821	0·8844	0·8851	1·070
0·4	0·8585	0·8716	0·8707	0·8643	1·198
0·5	0·9223	0·8581	0·8570	0·8567	1·342
0·6	0·8381	0·8357	0·8355	0·8356	1·554
0·7	0·8376	0·8377	0·8375	0·8374	1·548
0·8	0·8310	0·8140	0·8176	0·8147	1·776
0·9	0·8045	0·8029	0·8025	0·8022	1·910
1·0	0·8107	0·8090	0·8083	0·8082	1·848
1·05	0·8110	0·8098			
1·1	0·8078	0·8077	0·8075		1·862
1·2	0·9922	0·7929	0·7933		2·014
1·4	0·7529	0·7535	0·7538		2·458
1·6	0·7364	0·7359	0·7360		2·664
1·8	0·7415	0·7421	0·7422		2·600
2·0	0·7528	0·7517	0·7518		2·478
2·2	0·7608		0·7600		2·386
2·4	0·7622		0·7616		2·358
2·6	0·7571		0·7578		2·404
2·8	0·7472		0·7467		2·536
3·0	0·7330		0·7328		2·708
3·2	0·7162		0·7163		2·904
3·4	0·6982		0·6982		3·118
3·6	0·6794		0·6792		3·364
3·8	0·6598		0·6596		3·607
4·0	0·6409		0·6407		3·862
5·0	0·5511				
6·0	0·4780				
7·0	0·4198				

for the drop in average pressure amplitude expressed in decibels (db) where the value in decibels is given by db $= -20\log_{10} p_{\text{rms}}$. These last figures are useful for finding the correction necessary for the pressure measurements made between any two distances—a useful rule of thumb is to take the correction as being one decibel per r_o^2/λ. While the correction does depend upon βr_o this dependence becomes negligible for $\beta r_o \geqslant 100$ when $r_o^2/\lambda > 1$ and for $\beta r_o \geqslant 50$ when $r_o^2/\lambda > 2$. The decibel correction to pressure amplitude for $\beta r_o \geqslant 200$ is plotted in Fig. 2.7.

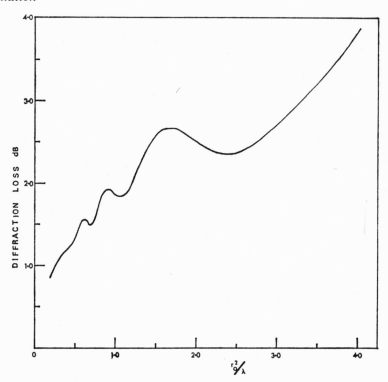

DIFFRACTION LOSS dB

Fig. 2.7 Decibel correction allowing for diffraction loss as a function of the distance from a circular piston expressed in units of r^2/λ for $\beta r_0 > 200$. (After Seki *et al* (2.3).)

2.4 Cylindrical Waves

Ultrasonic radiation usually takes the form of plane or quasi plane waves but we have found it necessary to consider spherical radiation in order to understand the operation of an apparently plane-wave generator such as a finite piston. For completeness we should also consider cylindrical waves, that is waves generated by an infinitely long cylinder whose radius is pulsating sinusoidally. In this case the radiated pressure will be a function of time and r only where r denotes the distance of a point in the radiation field measured from the cylinder axis. In this case the wave equation (equation 2.2) when transformed into cylindrical co-ordinates becomes:

$$\frac{\partial^2 p}{\partial t^2} = \frac{c^2}{r}\frac{\partial}{\partial r}\left(r\frac{\partial p}{\partial r}\right) \quad \cdot \quad \cdot \quad \cdot \quad \cdot \quad \textbf{(2.18)}$$

The general solution of equation (2.18) is:

$$p = A[J_o(\beta r) - jN_o(\beta r)]e^{j\omega t} \quad \cdot \quad \cdot \quad \cdot \quad \textbf{(2.19)}$$

where $\beta = \dfrac{2\pi}{\lambda}$, A is a constant and J_o and N_o are Bessel and Neumann functions respectively. While cylindrical waves are sinusoidal in time, if the cylinder radius pulsates sinusoidally they are sinusoidal in distance only at large distances as may be seen if we consider the limiting expressions for p:

$$\beta r \to 0, \ p \to \ -j\frac{2}{\pi} A \log_e (\beta r) . e^{j\omega t} \quad . \quad . \quad \textbf{(2.20a)}$$

$$\beta r \to \infty, \ p \to \ A\sqrt{\frac{2}{\pi \beta r}} e^{j(\omega t - \beta r) + j\frac{\pi}{4}} \quad . \quad . \quad \textbf{(2.20b)}$$

2.5 Scattering

If we have a region of space traversed by a beam of sound waves there will be a particular distribution of acoustic pressures throughout the space. If now we introduce a body into the path of the waves the pressure distribution will be modified (except in the trivial case in which the body has the same acoustic properties as the medium in which it is placed) because the body will scatter the incident radiation. We can regard the acoustic pressure at any point as being the sum of the pressure, which would exist in the absence of the body, and that due to the scattered wave radiated by the body. The mathematical analysis of scattering reduces to finding the appropriate distribution, strengths and relative phases of Huyghens sources required to represent the scattered radiation. This is by no means easy. As a simple example we consider a plane wave travelling to the right falling normally on to a plane sheet of material with infinite specific acoustic impedance. We know that the incident radiation will be totally reflected. We can represent this situation by postulating that the front surface of the reflecting sheet radiates two waves, one travelling to the right and one to the left. If the radiation travelling to the right has the same amplitude as, but is in antiphase to, the incident wave, these two will cancel out so that there is no radiation present to the right of the front surface of the sheet. This leaves just the two beams on the left of the sheet, the incident beam and the reflected beam.

The polar diagram of the radiation scattered by a body depends upon the shape of the body and its size relative to the wavelength of the incident radiation. We can distinguish three cases; first, when the body is very much larger than a wavelength the radiation is reflected more or less specularly so that the body casts an acoustic shadow, secondly, when the body is very much smaller than a wavelength the radiation is hardly scattered at all so that the incident radiation diffracts round it and, thirdly, when the body dimensions are of the same order of magnitude as the wavelength the polar diagram for the scattered wave becomes complicated and greatly dependent upon the body dimensions.

There are two cases of practical importance which are amenable to calculation; the rigid cylinder and rigid sphere. We shall merely quote approximate expressions for the scattered intensity, the exact expressions can be found in Morse[2.4].

2.5a The cylinder

The intensity $I_{r\varphi}$ of the scattered radiation at a distance r from the cylinder axis and in a direction which makes an angle φ with the incident radiation when measured in a plane containing the incident beam and which is perpendicular to the cylinder is given by:

$$I_{r\varphi} \approx I_0 \frac{\pi r_0}{8r}(\beta r_0)^3(1 - 2\cos\varphi)^2 \quad . \quad . \quad . \quad (2.21)$$

when $\beta r_0 \ll 1$, and where

$$I_0 = \text{incident intensity}$$
$$r_0 = \text{cylinder radius}$$
$$\beta = 2\pi/\lambda$$

Polar diagrams based on equation (2.21) are shown in Fig. 2.8.

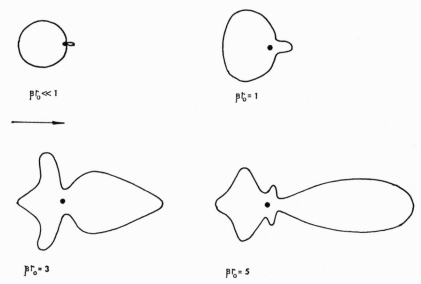

$\beta r_0 \ll 1$

$\beta r_0 = 1$

$\beta r_0 = 3$

$\beta r_0 = 5$

Fig. 2.8 Polar diagrams for the radiation scattered from a rigid cylinder of radius r_0.

The total power scattered per unit length, W_s, is a function of r_0/λ and is illustrated in Fig. 2.9. While the exact expression for the curve in Fig. 2.9 is somewhat complicated it reduces to simple forms for limiting values of βr_0, viz.:

$$\beta r_0 \ll 1, \quad W_s \approx 7.5(\beta r_0)^3 r_0 I_0 \quad . \quad . \quad . \quad (2.22a)$$

$$\beta r_0 \gg 1, \quad W_s \approx 4r_0 I_0 \quad . \quad . \quad . \quad . \quad (2.22b)$$

A useful measure of the effectiveness of a scatterer is its scattering cross-section which is defined as the ratio of the total power scattered to the incident intensity. Thus when $\beta r_0 \gg 1$ the scattering cross-section per unit length of the cylinder is $4r_0$. This result is apparently paradoxical since it implies that the total energy scattered by the cylinder is twice that actually incident upon it since the area presented to the incident energy is only $2r_0$ per unit length. We may resolve this paradox by noting that here we are dealing with a cylinder whose radius is large compared with a wavelength so that the cylinder will cast an acoustic shadow which

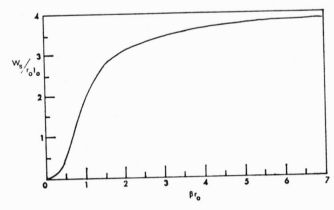

Fig. 2.9 Scattering power per unit length for a cylinder of radius r_0 as a function of βr_0.

arises from the destructive interference between the forward scattered energy and the incident energy. With a large cylinder this shadow will be fairly well defined and will have a cross-sectional area equal to that of the cylinder. Thus the power in the beam which destructively interferes must equal the incident power, that is $2r_0 I_0$ per unit length. This leaves an amount of power $2r_0 I_0$ per unit length to be reflected and this equals the incident power as we would expect.

If the cylinder is not completely rigid some energy will flow into it and possibly be absorbed. In this case we can define an absorption cross-section as the ratio of total power absorbed to the incident intensity.

2.5b The sphere

The intensity $I_{r\theta}$ of the radiation scattered from a sphere of radius r_0 at a distance r and in a direction which makes an angle θ to the incident radiation is given by:

$$\beta r_0 \ll 1, \; I_{r\theta} \approx 0\cdot 11 \; (\beta r_0)^4 \left(\frac{r_0}{r}\right)^2 (1 - 3\cos\theta)^2 \quad . \quad . \quad . \quad \textbf{(2.23a)}$$

$$\beta r_0 \gg 1, \; I_{r\theta} \approx \left(\frac{r_0}{2r}\right)^2 + \left(\frac{r_0}{2r}\right)^2 \cot^2\left(\frac{\theta}{2}\right) . \; J_1{}^2(\beta r_0 \sin\theta) \quad \textbf{(2.23b)}$$

The total scattered power W_s, as a function of βr_0, is shown in Fig.

2.10. As for the cylinder we have fairly simple expressions for limiting values, viz.:

$$\beta r_o \ll 1, \quad W_s \approx 5.6 \, (\beta r_o)^4 r_o^2 I_o \quad . \quad . \quad . \quad \text{(2.24a)}$$

$$\beta r_o \gg 1, \quad W_s \approx 2\pi r_o^2 I_o \quad . \quad . \quad . \quad . \quad \text{(2.24b)}$$

As with cylinders we notice that the scattering cross-section when $\beta r_o \gg 1$ is twice the cross-section of the sphere presented to the incident

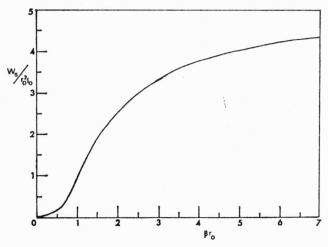

Fig. 2.10 Scattering power of a sphere of radius r_o as a function of βr_o.

radiation. When $\beta r_o \ll 1$ the ratio of scattered to incident power increases with the fourth power of βr_o which is the same relation as Rayleigh scattering for light.

2.6 Focusing Systems

Since acoustic waves have many properties similar to those of light waves in that they can be scattered, reflected and refracted we would expect to be able to construct lenses and mirrors analogous to optical systems and this in fact proves to be so.

Lenses are used to focus radiation travelling in fluids and are usually made from machined solids. As the velocity of sound in a solid is usually greater than in fluids a wave obliquely incident upon a solid surface is refracted away from the normal (cf. equations 1.54) in contrast to the behaviour of a light ray incident on to say, a glass plate from air. This means that whereas a concave lens is used optically in order to diverge a beam of light it will converge an acoustical beam. A typical ultrasonic lens is illustrated in Fig. 2.11. The focal length f of such a plano–concave lens, which is the usual type employed, is given by:

$$f = \frac{R}{1 - c_l/c_s} \quad . \quad . \quad . \quad . \quad \text{(2.25)}$$

where R denotes the radius of curvature of the curved surface and c_l and c_s the sound velocities in the liquid and solid respectively. Apart from machinability the material from which a lens is made should (*a*) have a similar specific acoustic impedance as the surrounding medium in order to reduce, if not eliminate reflections at the interfaces, (*b*) a different acoustic velocity from the surrounding medium and (*c*) as low an absorption as possible. Few combinations of liquid and solid satisfy these criteria and most lenses are made from plastics such as polystyrene and polymethylmethacrylate[2.5]. The former is better, particularly at high-power levels since its absorption is smaller than the latter's. Fluid lenses have been made with carbon tetrachloride contained in thin-walled enclosures[2.6]. If the wall is very thin such lenses reflect about 1 % of the incident energy if the surrounding fluid is water compared with about 12 % for polystyrene. Solid lenses, however, are easier to construct and are usually plano–concave in order to reduce the conversion to shear waves at the surfaces.

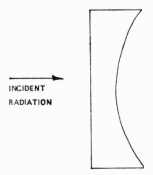

INCIDENT
RADIATION

Fig. 2.11 Cross-section through an ultrasonic focusing lens.

As well as lenses, reflectors can be used to focus acoustic radiation but, in contrast to lenses, the specific acoustic impedance of the reflector material must be as different as possible from the fluid. If the fluid is water, nickel and steel are the most suitable materials as each reflects about 89 % of the incident energy. A simple type of reflector would be a paraboloid which requires the generator of the plane waves to be in front of it. This is not always convenient and in this case an alternative configuration due to Barone[2.7] may be used. This is shown in Fig. 2.12. The incident plane wave is converted into a cylindrical wave by the 90° cone. This cylindrical wave is reflected by the paraboloid section where it is converted into a spherical wave converging to the focus F. The system must be designed so that the waves reflected by the parabolic section are not obstructed by the cone and in his paper Barone gives a technique for ensuring this.

A third technique for focusing radiation employs a curved generator. Certain ceramic materials are often used to generate ultrasonic waves and these may be formed during manufacture into a hemisphere.

Even though we may have a perfectly plane incident wave and a perfect lens or reflector the region of maximum intensity at the focus will not be infinitesimally small because diffraction effects still occur. The

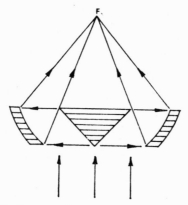

Fig. 2.12 Cross-section through Barone's reflector.

intensity I_z at a point, distant z from the system axis and in the plane passing through the focus is given by:

$$I_z = \frac{W}{(f\lambda)^2} \pi^2 R^4 \left[\frac{2J_1(x)}{x^2}\right]^2 \qquad \qquad (2.26)$$

where $x = \dfrac{2\pi z R}{f\lambda}$

and R = cross-sectional radius of incident beam
W = total incident acoustic power
f = focal length of lens or mirror
λ = wavelength of sound.

Equation (2.26) is similar to that for the intensity distribution in front of a circular piston $\left(\dfrac{z}{f} \approx \sin\theta\right)$ and like the latter it will have a zero when $x = 3.832$. About 84% of the total energy is contained within the main lobe which has a radius at the focus given by:

$$z_0 = 0.61\,\lambda f/R \qquad \qquad (2.27)$$

A convenient index of the concentrating power of a mirror or lens is the gain factor which is defined as the ratio of the peak pressure at the focus to the peak pressure in the plane wave. For a parabolic reflector the gain factor G is given by [2.8]:

$$G = 2\beta f \log_e(a^2/4f^2)\ (\lambda \leqslant 10a) \qquad \qquad (2.28)$$

where a denotes the radius of the mirror aperture. G has a maximum for a given aperture, when:

$$f \approx 0.18a \qquad . \quad . \quad . \quad . \quad \textbf{(2.29)}$$

and the maximum value is given by:

$$G_{max} \approx 4.6a/\lambda \qquad . \quad . \quad . \quad . \quad \textbf{(2.30)}$$

To give an order of magnitude for a perfect parabolic mirror the maximum theoretical gain at 1 mc/s in water for a mirror of 6 cm aperture is 92.

2.7 Radiation Pressure

When radiation falls upon an interface separating two media such that there is a difference in energy density between the two media the interface experiences a steady unidirectional force which is called radiation pressure. In order to show this and calculate the magnitude of this radiation pressure we consider the system shown in Fig. 2.13. Plane waves

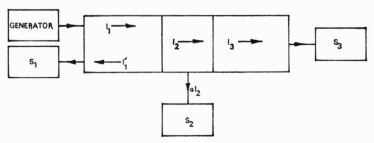

Fig. 2.13 Hypothetical system for the derivation of expressions for radiation pressure.

travel from the generator through medium 1 and are normally incident upon the front surface of medium 2 where they are partially reflected. The partially reflected waves travel to the left and are eventually absorbed completely by the 'sump' S_1 thereby ensuring that the incident intensity is simply that due to the waves coming from the generator and is not complicated by further reflected waves. The energy passing into medium 2 is partially absorbed and we represent this by assuming that the absorbed energy is abstracted by sump S_2. The unabsorbed energy in medium 2 passes without reflection into medium 3 which implies that the specific acoustic impedances of media 2 and 3 are identical. However, in order to bring out one facet of radiation pressure we assume that while the specific acoustic impedances of media 2 and 3 are identical their acoustic velocities differ. All the energy flowing into medium 3 is absorbed in the sump S_3. We now proceed to calculate the increase in total energy of the system, excluding the generator, during a small interval of time δt during which the piston formed by medium 2 travels towards the generator with a velocity v. If this turns out to be greater than the energy input from the generator work will have been done in moving the piston implying that it will have been moved against the pressure which will be the radiation pressure. We will denote quantities associated with medium 1 by a suffix 1, medium 2 by a suffix 2 and medium 3 by a suffix 3, intensities by I, (the reflected intensity in medium 1 by I_1'), angular frequency by ω, the fraction of the energy passing into medium 2 which is absorbed by α and $\rho_2 c_2 / \rho_1 c_1$ by m. Furthermore we assume that the cross-sectional area of the system is unity. Because of the Doppler effect the angular frequencies in different parts of the system will differ and from equations (1.69) and (1.70) we have:

$$\omega_1' = \omega_1 \frac{1 + v/c_1}{1 - v/c_1} \qquad \ldots \quad (2.31a)$$

$$\omega_2 = \omega_1 \left(1 + \frac{v}{c_1}\right) \qquad \ldots \quad (2.31b)$$

$$\omega_3 = \frac{\omega_2}{1 + \dfrac{v}{c_3}} = \omega_1 \frac{1 + \dfrac{v}{c_1}}{1 + \dfrac{v}{c_3}} \qquad \ldots \quad (2.31c)$$

Denoting the particle displacement amplitude by ξ we have from equations (1.37) and (1.38) that:

$$I_1' = \frac{\rho_1 c_1}{2}(\omega_1')^2 \xi_1^2 \left(\frac{m-1}{m+1}\right)^2 = \frac{\rho_1 c_1}{2}\omega_1^2 \xi_1^2 \left[\frac{1 + \dfrac{v}{c_1}}{1 - \dfrac{v}{c_1}}\right]^2 \left(\frac{m-1}{m+1}\right)^2$$

$$(2.32)$$

From equation (1.59) the intensity I_2 passing into medium 2 is given by:

$$I_2 = \frac{\rho_2 c_2}{2}\omega_2^2 \xi_1^2 \left(\frac{2}{m+1}\right)^2 = I_1 m \left(\frac{2}{m+1}\right)^2 \left(1 + \frac{v}{c_1}\right)^2 \quad (2.33)$$

Similarly the intensity I_3 passing into medium 3 is given by:

$$I_3 = I_1(1 - \alpha)m \left(\frac{2}{m+1}\right)^2 \left[\frac{1 + \dfrac{v}{c_1}}{1 + \dfrac{v}{c_3}}\right]^2 \qquad \ldots \quad (2.34)$$

According to equation (1.37) the energy density in a plane wave is given by I/c thus we can now draw up a table of the change in energy during δt for the various regions of our system due to either absorption or a change in volume.

Inserting now expressions for I_1', I_2 and I_3 from equations (2.32), (2.33) and (2.34) into the terms in table 2.3 we can find the total increase

TABLE 2.3. *Increase in energy during δt for various regions of system*

Region	Increase in Energy
Sump 1	$I_1'\delta t$
Medium 1	$-[I_1'v/c + I_1 v/c]\delta t$
Sump 2	$\alpha I_2 \delta t$
Medium 3	$I_3(v/c_3)\,\delta t$
Sump 3	$I_3\delta t$

in energy of the system which must equal the energy input from the generator of I_1 . δt plus the work done in moving the piston of medium 2 against a radiation pressure P'_{rad}, thus we have:

$$I_1\left[1 + \frac{v}{c_1} + \frac{v}{c_1}\left(\frac{m-1}{m+1}\right)^2 - \left(\frac{4m}{m+1}\right)^2\left(1 + \frac{2v}{c_1}\right)(1 - \alpha)\frac{v}{c_3}\right]\delta t$$
$$= I_1\delta t + P'_{\text{rad}}\, v\, \delta t \quad . \quad . \quad . \quad . \quad (2.35)$$

Whence:

$$P'_{\text{rad}} = I_1\left\{\frac{1}{c_1}\left[1 + \left(\frac{m-1}{m+1}\right)^2\right] - \frac{4m}{(m+1)^2}\left(1 + \frac{2v}{c_1}\right)(1 - \alpha)\frac{1}{c_3}\right\}$$
$$(2.36)$$

If we now assume the velocity v to fall to zero a radiation pressure P_{rad} still exists and this is the value which is normally taken, viz.:

$$P_{\text{rad}} = I_1\left[\frac{1}{c_1}(1 + R) - (1 - R)(1 - \alpha)\frac{1}{c_3}\right] \quad . \quad (2.37)$$

where $R = \left(\frac{m-1}{m+1}\right)^2$ which is the reflection coefficient for intensities.

While equation (2.37) is a general expression for the pressure acting on medium 2 in our system we can obtain a better understanding of radiation pressure in general by taking a few special cases.

Case 1. Perfect reflector, $R = 1$.

$$P_{\text{rad}} = \frac{2I_1}{c_1} = 2E_1 \quad . \quad . \quad . \quad . \quad (2.38)$$

where E_1 denotes the energy density of the incident wave.

Case 2. Perfect absorber, $R = 0$, $\alpha = 1$.

$$P_{\text{rad}} = \frac{I_1}{c_1} = E_1 \quad . \quad . \quad . \quad . \quad (2.39)$$

Case 3. Zero absorption and reflection in medium 2.

$$P_{\text{rad}} = \frac{I_1}{c_1} - \frac{I_1}{c_3} = E_1 - E_3 \quad (I_1 = I_3) \quad . \quad . \quad \textbf{(2.40)}$$

Case 4. Zero reflection but some absorption in medium 2.

$$P_{\text{rad}} = \frac{I_1}{c_1} - \frac{I_1(1 - \alpha)}{c_3} = \frac{I_1}{c_1} - \frac{I_3}{c_3} = E_1 - E_3 \ . \quad \textbf{(2.41)}$$

These four special cases all show that the radiation pressure acting on a plane equals the difference in energy density between the two sides of the plane and acts in the direction of decreasing energy density.

Throughout the derivation of radiation pressure we have assumed that the waves are perfectly sinusoidal and the medium is linear so that radiation pressure is not due to non-linearities in wavemotion as is sometimes assumed.

2.8 Streaming

We saw in the last section that a radiation pressure is exerted on an element of, say, fluid when there is a difference in energy density between the ends of the element. If a beam of ultrasonic waves travels through a fluid which absorbs ultrasonic energy there will be an ultrasonic intensity gradient set up with which will be associated a corresponding

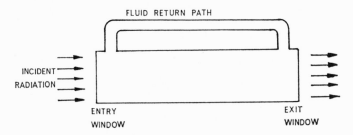

Fig. 2.14 Simplified system for streaming studies.

energy density gradient which implies that there will be a radiation pressure exerted on a small element of the fluid. This pressure will cause the fluid to flow in the direction of propagation of the radiation and return by some path where the ultrasonic intensity is lower or non-existent. Such a flow is known as streaming.

In order to derive an expression for streaming we consider a system as illustrated in Fig. 2.14 in which a beam of ultrasonic waves enters, and completely fills a cylinder of radius r_0, through a transparent window and leaves by another window. The cylinder contains a fluid of absorption coefficient α and acoustic velocity c. Attached to the cylinder is another which provides a return path for the fluid. If the pressure acting on the column of fluid is P, the volume flowing per second, Q, is given by Poiseuille's formula as:

$$Q = \frac{\pi P r_0{}^4}{8\eta l} \quad \cdots \quad \cdots \quad (2.42)$$

where η denotes the fluid viscosity and l the fluid column length.

If the intensity of the radiation entering the column is I_0 the difference in energy density between the ends of the column, which will equal the radiation pressure acting on the column, is given by:

$$\frac{I_0}{c}(1 - e^{-2\alpha l}) = P \quad \cdots \quad \cdots \quad (2.43)$$

If now $2\alpha l \ll 1$, as is usually the case, at least for liquids with a lowish absorption coefficient, we have on combining equations (2.42) and 2.43):

$$Q = \frac{\pi r_0{}^4 I_0 \alpha}{\eta c} \quad \cdots \quad \cdots \quad (2.44)$$

If we can measure Q and know I_0 and η we can find the absorption coefficient for the fluid. This technique will be considered later when we deal with measuring techniques.

2.9 The Measurement of Radiation Intensity

It is often necessary, particularly for medical applications, to be able to measure the intensity of ultrasonic radiation. At the present time three methods are available for measuring intensity; the first measures the power output from the transducer by measuring the electrical power input and the transducer efficiency (cf. section 8.6 and reference 8.8) while the second and third methods measure the radiated power directly by calorimetric and radiation pressure techniques respectively.

The calorimetric method usually employs an absorber which completely absorbs the incident radiation and the resulting temperature rise is monitored and converted into an acoustic power indication by means of a calibration technique in which the amount of electrical power required to raise the temperature by the same amount is measured. One such system, developed by Wells and his co-workers[2.9], uses a sphere filled with carbon tetrachloride into which the radiation is passed tangentially thereby ensuring that all the incident radiation is absorbed within the sphere and none reradiated. The resulting temperature rise is monitored by a thermocouple within the sphere. An alternative technique due to Mikhailov[2.10] uses a horn-shaped Dewar flask filled with an absorbing liquid. The resulting temperature rise causes the liquid to expand and this expansion is monitored by the rise in liquid level in an attached capillary tube. The absorber contains its own internal electrical heating element for calibration purposes.

Fry and Fry[2.11] developed an alternative calorimetric technique which does not require the absorber to absorb the incident radiation completely. In their system, shown in Fig. 2.15, a very fine thermocouple wire is

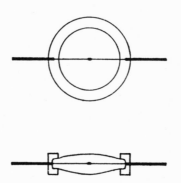

Fig. 2.15 Thermocouple probe. (After Fry and Fry (2.11).)

stretched across the path of the radiation within a small container filled with an absorbing liquid whose impedance matches that of the liquid surrounding the container. For measurements in water they found castor oil to be suitable. The container faces are made from thin polythene sheet in order that they obstruct the passage of the radiation as little as possible. When the transducer is switched on the thermocouple registers a rapid temperature rise followed by a slow linear rise as shown in Fig. 2.16. This initial rise is due to the viscous motion of the castor oil

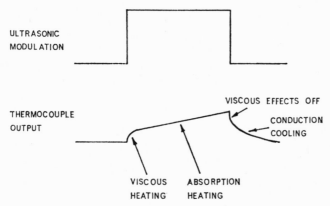

Fig. 2.16 Voltage waveforms obtained from Fry and Fry's thermocouple probe.

past the thermocouple and which rapidly attains its equilibrium value. The succeeding slower temperature rise is caused by the absorption of the radiation within the castor oil. If this slower rise is extrapolated back to zero time to give $(dT/dt)_{t\,=\,0}$, Fry and Fry showed that the intensity I is given by:

$$I = \left(\frac{\rho}{2\alpha}\right) C \left(\frac{dT}{dt}\right)_{t\,=\,0}$$

where α denotes the absorption coefficient for castor oil, ρ its density and C its heat capacity per unit mass at constant pressure. The authors claim that this device has the advantages of being physically small, stable, insensitive to stray electrical fields and is sufficiently accurate to be regarded as a primary standard. Its disadvantages are that it has a low sensitivity and is only accurate if the radiation is a pure sine wave since the presence of harmonics will cause uncertainty in the value to be ascribed to the absorption coefficient.

We saw in section 2.7 that whenever an ultrasonic energy-density difference exists across a boundary that boundary experiences a unidirectional force known as radiation pressure. This pressure can be used to measure the intensity of the radiation if the reflecting characteristics of the boundary and the velocity of the radiation in the medium are

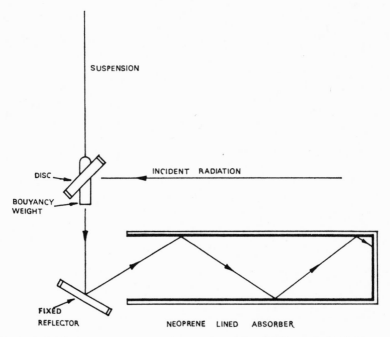

Fig. 2.17 Radiation pressure metering system. (After Wells *et al.* (2.12).)

known. In practice two types of boundary are used; totally absorbing and totally reflecting with the latter being the more common. A typical radiation pressure meter is that shown in Fig. 2.17 due to Wells, Bullen and Freundlich [2.12]. A hollow aluminium disc is suspended by two phosphor bronze wires so that it makes an angle of 45° to the direction of travel of the incident radiation. The reflected radiation travels down-wards until it meets another, fixed, reflector which diverts it into a tube lined with neoprene which absorbs 99 % of the radiation thereby elimi-nating standing waves in front of the disc which would lead to errors. From a knowledge of the dimensions of the system the deflection of the disc can easily be converted into radiation pressure and thence into the radiation intensity. Wells, Bullen and Freundlich claim an accuracy of + 3 % at a power level of 2 milliwatts after taking into account various sources of error the most important of which being the weight of the suspension wires.

As an alternative to a plane-reflecting surface a spherical one can be used with the advantage that the incident radiation is scattered in all directions and thereby minimising the chances of setting up a standing-wave system. According to Fox[2.13] the force F acting on the sphere due to radiation pressure, expressed in Newtons, placed in a radiation field of intensity I watt/m^2 is given by:

$$F = \pi a^2(I/c)[1 - 0.719(2\pi a/\lambda)^{-4/3}]$$

where a denotes the sphere radius, c the velocity of sound in the medium and λ its wavelength. This expression is only valid if $2\pi a/\lambda \gg 1$.

In making radiation-pressure measurements it is important to elimi-nate any additional force acting upon the reflector which may arise from streaming by placing a very thin membrane just in front of the reflector.

References

2.1 Williams, A. O., *J. Acoust. Soc. Amer.*, **23**, 1 (1951).

2.2 King, L. V., *Proc. Roy. Soc.*, **147A**, 212 (1934).

2.3 Seki, H., Granato, A. and Truell, R., *J. Acoust. Soc. Amer.*, **28**, 230 (1956).

2.4 Morse, P., *Vibration and Sound*, p. 346. Wiley (1948).

2.5 Wall, P. D., Tucker, D., Fry, F. J. and Mosberg, W. A., *J. Acoust. Soc. Amer.*, **25**, 281 (1953).

2.6 Pohlman, R., *Z. Phys.*, **113**, 697 (1939).

2.7 Barone, A., *Acustica*, **2**, 221 (1952).

2.8 Horton, C. W. and Nolle, F. C., *J. Acoust. Soc. Amer.*, **22**, 855 (1950).

2.9 Wells, P. N. T., Bullen, M. A., Follett, D. H., Freundlich, H. F. and James, J. A., *Ultrasonics*, **1**, 106 (1963).

2.10 Mikhailov, I. G., *Ultrasonics*, **2**, 129 (1964).

2.11 Fry, W. J. and Fry, R. B., *J. Acoust. Soc. Amer.*, **26**, 294 (1954).

2.12 Wells, P. N. T., Bullen M. A. and Freundlich, H. F., *Ultrasonics*, **2**, 124 (1964).

2.13 Fox, F. E., *J. Acoust. Soc. Amer.*, **12**, 147 (1940).

Transducers

In the previous two chapters we developed the basic theory for acoustic waves, but, apart from radiation patterns, we have paid little attention to the wave generator. Generators, or transducers as they are more usually termed, convert energy in one form or another into acoustic energy and can be roughly divided into two types—mechanical and electromechanical.

3.1 Mechanical Transducers

Mechanical transducers convert the kinetic energy of a stream of fluid into acoustic energy and are either whistles or sirens.

3.1a Galton's whistle

The principle of Galton's whistle[3.1] is shown diagrammatically in Fig. 3.1. A stream of compressed air is pumped through an annular orifice

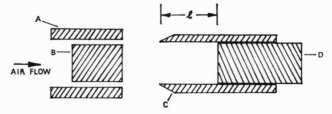

Fig. 3.1 Cross-section through Galton's whistle. *A* and *B*—concentric cylinders, *C*—wedge, *D*—plunger.

formed by the gap between two concentric cylinders *A* and *B*, after which it impinges on to the wedge-shaped edge of a resonant cavity formed by a cylinder *C* and plunger *D*. The air stream sets the cylinder end into oscillation at a frequency f governed by the dimensions of the cavity according to equation (3.1).

$$f = \frac{c}{4(l + k)} \qquad \cdots \cdots \quad (3.1)$$

where c denotes the velocity in mm/sec, l the length of the cavity in mm and k a constant which depends upon the gas pressure and ranges from about 4·5 to 7·5 with a typical value under normal operating conditions of about 6. The frequency range of this type of transducer is from about

3 to 50 kc/s. While the efficiency is of the order of a few per cent only, several watts of acoustic power can be generated.

3.1b Hartmann's whistle

Hartmann's whistle[3.2] is similar to Galton's but works on a different principle. Fig. 3.2 shows a schematic cross-section of the whistle. A jet

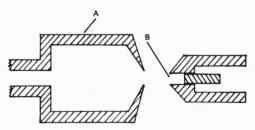

Fig. 3.2 Cross-section through Hartmann's whistle. *A*—cylinder, *B*—resonant cavity.

of air flows through the cylinder *A* which has a reduced cross-sectional area where the air exits. At the exit the air velocity increases and provided that the air pressure within *A* exceeds ambient by about 0·9 atmos the exit velocity exceeds that of sound giving rise to a shock wave a short distance out from the end of the cylinder. At this point the shock wave is unstable and causes the air within the cavity *B* to oscillate at a frequency *f* given by:

$$f = \frac{c}{4(l + 0 \cdot 3d)} \quad \cdot \quad \cdot \quad \cdot \quad \cdot \quad \cdot \quad (3.2)$$

Again the efficiency is low but output powers of up to 100 watts can be generated.

3.1c Jet-edge systems

Jet-edge generators[3.3] are suitable for both gases and liquids and provide a stable and simple means for generating fairly high powers suitable

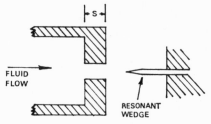

Fig. 3.3 Cross-section through a jet-edge system.

for bulk processing of fluids by ultrasonic energy such as manufacturing emulsions and dispersions. The principle under which these generators work is shown in Fig. 3.3. A stream of fluid is pumped through a circular hole in a plate of thickness *s*. Because of viscous forces at the fluid

C

solid boundary, vortices are set up which travel with a velocity v, say, until they reach the far side of the plate where they radiate a pressure pulse in all directions. The pressure pulse radiated by one such vortex will travel back to the other side of the plate where it will provoke another vortex to form. The time interval separating the vortices will thus be about s/v so that the hole in the plate will effectively radiate pressure pulses at a frequency of about v/s. The stability of the system can be improved by placing a suitably resonant wedge-ended plate near the hole. This type of generator can produce several watts of acoustic power at frequencies ranging up to about 10 kc/s.

3.1d Sirens

A siren generates acoustic power by rhythmically interrupting a stream of gas. In a typical siren[3.4] a stream of air is passed through a series of holes situated around the circumference of an annular chamber. These holes are alternately opened and closed by a tooth-edged disc rotating so that the teeth pass across the holes. Sirens can generate several hundred watts of acoustic power at frequencies up to about 20 kc/s and at efficiencies which can be as high as 70%.

3.2 Electromechanical Transducers

Electromechanical transducers convert electrical into acoustic energy There are five types of such transducers in current use: moving coil, electrostatic, piezoelectric, ferroelectric and magnetostrictive types. The moving-coil transducer[3.5] is similar to the moving-coil loudspeaker and finds its greatest application as a vibrator for acceleration testing of components. The ribbon type of moving-coil transducer, in which a thin ribbon of aluminium, situated between the poles of a magnet, acts like a partial turn of the coil of a moving-coil unit, has been used to generate ultrasonic waves in studies of the ultrasonic absorption of gases. The electrostatic transducer depends on the fact that the opposing plates of a parallel plate capacitor experience a force when a potential difference is applied between them. Its main use is as a high-quality audio loudspeaker but it has been used to excite ultrasonic vibrations in solids[3.6]. The remaining three types of transducer are the commonest in use and we will deal with these in some detail.

3.3 Piezo- and Ferro-electricity

Certain polar molecules form crystals which lack centres of symmetry. If a plate cut from such a crystal is mechanically deformed a voltage develops between two faces of the plate and this phenomenon is known as the direct piezoelectric effect. Conversely, if a voltage is applied between the two faces the plate will deform because of the inverse piezoelectric effect.

As an example of the mechanism underlying piezoelectricity we consider one of the most widely used naturally occurring piezoelectric materials, quartz, one unit cell of which is shown in Fig. 3.4 situated between two parallel planes. In the undeformed state, shown in Fig. 3.4a, the centres of gravity of the positive silicon ions and the negative oxygen ions coincide and, as the ionic charges are equal, there is no

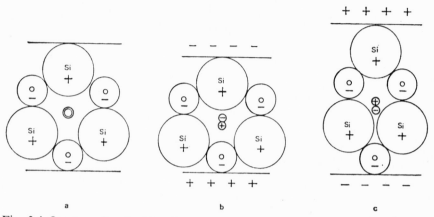

Fig. 3.4 Quartz unit cell. (*a*) cell unstressed; (*b*) cell in compression; (*c*) cell in extension.

resultant dipole moment so that the effective charge distribution across the parallel planes is zero. However, if the unit cell is deformed, by, for example, compression as in Fig. 3.4b or extension as in Fig. 3.4c, the centres of gravity of the silicon and oxygen ions no longer coincide leading to an apparent charge distribution appearing at the parallel planes. Deformations apart from compression or extension in the directions shown in Fig. 3.4 will separate the centres of gravity and lead to charges appearing at the parallel planes. The inverse piezoelectric effect arises from the fact that if the unit cell is placed in an electric field the centres of gravity will tend to move in opposite directions causing the cell to deform. If we have a quartz plate cut from a crystal in such a way that the strain is parallel to the applied electric field and we apply an alternating electric field the plate will expand and contract with the same frequency as the field; that is it will behave as a piston and can thus act as a generator of acoustic waves.

While only a few materials exhibit piezoelectricity, at least to a marked degree, all dielectrics are electrostrictive, that is they show an increase in length parallel to an applied electric field. Usually the magnitude of this effect is small but a few materials, such as certain titanates and zirconates, show a large effect in which case they are said to be ferroelectric. Normally these materials are made in the form of a ceramic composed of a large number of randomly orientated polarised domains. If such a material is heated to above its Curie temperature and then cooled in the presence of a strong electric field the directions of polarisation of the domains line up with the field and remain lined up after the field is removed when the material is cool. Such a material now possesses a permanent electric dipole and behaves to all intents and purposes as if it were piezoelectric obeying similar equations to those for a naturally piezoelectric material. This being so we have no further need to distinguish between the two and we shall use the term piezoelectricity to describe the properties of both truly piezoelectric and ferroelectric materials.

Since piezoelectric materials are anisotropic the equations relating electrical to mechanical properties will necessarily involve a large number of coefficients in equations similar to the general equations of elasticity (equations 1.9). In practice six different types of coefficient are employed in order to describe piezoelectric materials although not all these are independent. We have the elastic stiffness coefficients c_{ij}, where the subscripts i and j have the same meaning as in chapter 1, the absolute permittivity ε_{ij}, two piezoelectric strain coefficients, d_{ij} and g_{ij}, and two piezoelectric stress coefficients e_{ij} and h_{ij}. The first subscript denotes the vector component of the electric field and the second subscript the component of stress or strain. Thus if we take the Z direction as the direction of the principal axis in the case of quartz and as the direction of polarisation of our ferroelectric and we apply an electric field parallel to Z the subscripts describing a stress or strain in the Z direction are both 3. If the stress or strain however occurs in the X direction with the field still in the Z direction, i still equals 3 but j is now 1.

Strictly speaking the equations for a piezoelectric material should be written either in a form similar to equation (1.9) or else in tensor notation. However, in order to simplify matters we assume that when a field is applied parallel to one of the axes the stress or strain is also in the same direction. For quartz this means that the field is applied parallel to the X axis but for ferroelectrics the field is parallel to the Z axis. As we have four possible piezoelectric coefficients we will have a total of eight equations, four for the direct piezoelectric effect and four for the inverse effect, relating the two electrical quantities, displacement D and field strength E, to two mechanical quantities, strain S and tension T.

TABLE 3.1. *Piezoelectric equations*

Piezoelectric Constant	Direct Effect	Inverse Effect
d	$D = \varepsilon^T E + dT$	$S = \dfrac{1}{c^E}T + dE$
e	$D = \varepsilon^S E + eS$	$T = c^E S - eE$
g	$E = \dfrac{1}{\varepsilon^T}D - gT$	$S = \dfrac{1}{c^D}T + gD$
h	$E = \dfrac{1}{\varepsilon^S}D - hs$	$T = c^D S - hD$

The equations are given in table 3.1 in which the superscripts denote the quantities held constant during the measurement.

The equations in table 3.1 enable us to define the four piezoelectric constants in the forms given in table 3.2:

TABLE 3.2. Definitions of piezoelectric constants

Constant	Definition	M.K.S Units
d	Charge density developed / Applied mechanical stress	Coulomb/Newton
	Strain developed / Applied field	Metre/Volt
e	Charge density developed / Applied mechanical strain	Coulomb/Metre²
	Stress developed / Applied field	Newton/Volt/Metre
g	Field developed / Applied mechanical stress	Volt/Metre/Newton
	Strain developed / Applied charge density	Metre²/Coulomb
h	Field developed / Applied mechanical strain	Volt/Metre
	Stress developed / Applied charge density	Newton/Coulomb

As we have previously mentioned the four piezoelectric constants are not independent and they are, in fact, related by the following equations:

$$e_{ij} = \sum_{h} c_{jh} \, d_{ih} \quad \ldots \quad \ldots \quad \textbf{(3.3a)}$$

$$h_{ij} = e_{ij}/\varepsilon_{ij}{}^{S} \quad \ldots \quad \ldots \quad \textbf{(3.3b)}$$

$$g_{ij} = d_{ij}/\varepsilon_{ij}{}^{T} \quad \ldots \quad \ldots \quad \textbf{(3.3c)}$$

3.4 Equivalent Circuit of a Piezoelectric Transducer

A piezoelectric transducer is driven from an electrical generator and therefore it will be convenient to be able to represent the transducer by means of an electrical circuit. In order to facilitate the derivation of a suitable circuit we shall restrict our analysis to the case of a flat disc of piezoelectric material, of thickness l and single surface area a, with electrodes attached to its plane surfaces in order to provide the energising field, vibrating in a thickness mode. That is we assume that the only dimension to change under the influence of the electric field is the thickness l. This restriction to a thickness mode is not strictly valid since there will be coupling to other modes, such as radial ones, because of the effect of Poisson's ratio. However, provided that the resonant frequencies of the other modes are well away from the frequency at which the transducer is being driven, their effects may be ignored.

If we assume the disc to be placed in a co-ordinate system with one face at $x = 0$ and the other at $x = l$ and that the positive direction of the applied field is in the positive x direction we can use the last two equations in table 3.1 to write:

$$T_x = c_x \frac{\partial \xi_x}{\partial x} - hD_x \qquad . \quad . \quad . \quad . \quad (3.4)$$

$$E_x = -h\frac{\partial \xi_x}{\partial x} + D_x/\varepsilon_x \qquad . \quad . \quad . \quad (3{\cdot}5)$$

where ξ_x denotes the displacement at a point x within the transducer, c_x the elastic constant in the x direction and ε_x the absolute permittivity of the transducer material in the x direction ignoring now the distinction between ε^S and ε^T and c^E and c^D.

If we assume that the applied field is sinusoidal the disc will expand and contract sinusoidally which is equivalent to stating that elastic waves are travelling backwards and forwards inside the disc. In other words the transducer is behaving as an acoustic transmission line. Thus equation (1.28) enables us to write down an expression for the displacement within the disc as follows:

$$\xi_x = [Ae^{-j\beta x} + Be^{j\beta x}]e^{j\omega t} \qquad . \quad . \quad . \quad (3.6)$$

where A and B are constants determined by the boundary conditions at $x = 0$ and $x = l$. If we differentiate equation (3.6) with respect to x and substitute into equation (3.4) we obtain, after noting that $\beta c_x = Z_c \omega/a$, where Z_c denotes the acoustic impedance, $\rho_c ca$, of the transducer material:

$$-(T_x + hD_x) = [(j\omega Z_c A)e^{-j\beta x} + (-j\omega Z_c B)e^{j\beta x}]e^{j\omega t} \quad (3.7)$$

Also since there is no free charge within the transducer we have that:

$$D_x = D_o = Q/a \qquad . \quad . \quad . \quad . \quad (3.8)$$

where Q denotes the charge on one of the transducer plane surfaces. Furthermore, the total tensile force on a plane at x is given by $F_x = T_x a$, thus after substituting equation (3.8) into equation (3.7) we obtain:

$$-(F_x + hQ) = [(j\omega Z_c A)e^{-j\beta x} + (-j\omega Z_c B)e^{j\beta x}]e^{j\omega t} \quad (3.9)$$

Also if we differentiate equation (3.6) with respect to time we find:

$$\frac{\partial \xi_x}{\partial t} = \frac{1}{Z_c} [(j\omega Z_c A)e^{-j\beta x} - (-j\omega Z_c B)e^{j\beta x}]e^{j\omega t} . \quad (3.10)$$

Equations (3.9) and (3.10) are formally identical with those of an electrical transmission line of characteristic impedance Z_c and with the term $\frac{\partial \xi_x}{\partial t}$ analogous to current and $-(F_x + hQ)$ analogous to voltage. We may, therefore, represent the transducer as in Fig. 3.5.

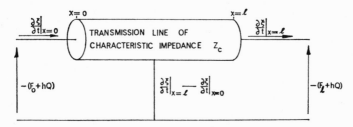

Fig. 3.5 Piezoelectric transducer equivalent circuit corresponding to equation (3.10).

The term hQ is common to both $x = 0$ and $x = l$ so we may redraw Fig. 3.5 as Fig. 3.6 in which we have also reversed the directions of F_x

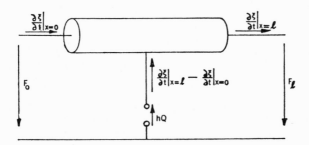

Fig. 3.6 Modified version of Fig. 3.5 with term hQ inserted into common lead.

and F_o in order to remove the negative sign. Our next step is to relate the force hQ to the voltage applied across the disc electrodes. Denoting the potential at x by V_x we have:

$$E_x = -\frac{\partial V_x}{\partial x} \qquad \cdots \qquad (3.11)$$

Combining equations (3.5), (3.8) and (3.11) and being careful about the sign of the potential difference between the electrodes, V, we find eventually:

$$V = \frac{Ql}{a\varepsilon} - h(\xi_l - \xi_o) \qquad \cdots \qquad (3.12)$$

Now we can write $\frac{a\varepsilon}{l} = C_o$ where C_o is the static capacitance of the disc when we regard the piezoelectric material simply as an insulator. Also since we assume sinusoidal motion we have $\xi_x = \frac{1}{j\omega}\frac{\partial \xi_x}{\partial t}$. Thus we can now write equation (3.12) as follows:

$$hQ = hC_oV + \frac{h^2C_o}{j\omega}\left[\frac{\partial \xi_l}{\partial t} - \frac{\partial \xi_o}{\partial t}\right] \qquad . \quad . \quad (3.13)$$

Since the term hQ denotes a force both terms on the right-hand side of equation (3.13) must also be forces. Taking the term hC_oV first, we may regard it as representing the force generated by an applied voltage V (this follows directly from the second definition of h in table 3.2). Thus the factor hC_o may be regarded as a transformation factor relating force to voltage and we may represent it diagramatically by a voltage-force 'transformer' of turns ratio hC_o for which when a voltage is applied across the primary a force appears across the secondary and vice versa. The total driving force hQ is the sum of the force across the transformer secondary and a force developed across some element through which flows a mechanical velocity (analogous to current) of $\left(\frac{\partial \xi_l}{\partial t} - \frac{\partial \xi_o}{\partial t}\right)$. Thus the quantity $h^2C_o/j\omega$ may be regarded as a mechanical impedance. This impedance, however, must be negative since the total force across it and the secondary of the transformer is greater than that across the secondary alone. Therefore we may regard the impedance as being provided by a mechanical capacitance of value $-1/h^2C_o$.

While our resulting equivalent circuit is correct on the mechanical side of the transformer it is not yet correct on the electrical side. The electrical current, i, flowing into the transducer is equal to $j\omega Q$, which, from equation (3.13) is given by:

$$i = j\omega C_oV + hC_o\left(\frac{\partial \xi_i}{\partial t} - \frac{\partial \xi_o}{\partial t}\right) \qquad . \quad . \quad . \quad (3.14)$$

The second term in equation (3.14) represents an electrical current flowing into the transformer primary because of the mechanical 'current' flowing out of the secondary while the first term represents a current

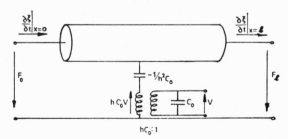

Fig. 3.7 Complete equivalent circuit for a piezoelectric transducer.

flowing into a capacitor of value C_0. Thus the complete equivalent circuit of the transducer is that given in Fig. 3.7.

The transmission line equivalent circuit of Fig. 3.7 is most useful for transient analysis of transducer behaviour but for steady state analysis it is more convenient if the transmission line is represented by its lumped parameter T equivalent circuit[3.7]. When this is done the final equivalent circuit for a transducer (assumed to be lossless) is that given in Fig. 3.8

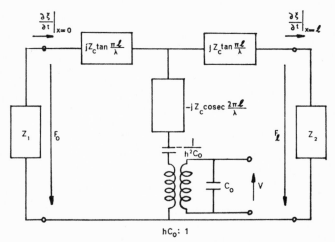

Fig. 3.8 Lumped parameter complete equivalent circuit for a piezoelectric transducer.

in which we have added Z_1 and Z_2 to represent the impedances of the media in contact with the two plane surfaces of the transducer.

3.5 Power Output from a Transducer

When a transducer is used to generate acoustic waves we are usually interested in the operating conditions under which the output is a maximum. Denoting the reactance of the negative mechanical capacitance by jX_c, the velocity at $x = l$ is given by:

$$\frac{\partial \xi_l}{\partial t} = \frac{2hC_0V\left[Z_1 + jZ_c \tan \frac{\pi l}{\lambda}\right] \tan \frac{\pi l}{\lambda}}{2(Z_1Z_2 + Z_c^2) \tan \frac{\pi l}{\lambda} - 4Z_cX_c \tan^2 \frac{\pi l}{\lambda} + j(Z_1 + Z_2)\left[2X_c \tan \frac{\pi l}{\lambda} + Z_c \tan^2 \frac{\pi l}{\lambda} - Z_c\right]} \quad \textbf{(3.15)}$$

The condition for maximum power output into Z_2 (Fig. 3.8) depends upon whether or not Z_1 is infinite. If Z_1 is infinite the condition for maximum power output is:

$$\tan \frac{\pi l}{\lambda} = 1 - \frac{1}{1 + 1 \cdot 5 \dfrac{Z_c}{X_c}} \quad \cdot \quad \cdot \quad \cdot \quad \cdot \quad \textbf{(3.16)}$$

For quartz the term Z_c/X_c is of the order of 300 so that for all practical purposes we may write:

$$\tan \frac{\pi l}{\lambda} \approx 1 \quad . \quad . \quad . \quad . \quad . \quad . \quad (3.17)$$

that is:

$$\frac{\pi l}{\lambda} = (2n + 1)\frac{\pi}{4}$$

or:

$$l = (2n + 1)\frac{\lambda}{4} \quad . \quad . \quad . \quad . \quad . \quad (3.18)$$

where $n = 0, 1, 2, 3, 4 \ldots \ldots \ldots$
Under this condition when the transducer is backed by an infinite acoustic impedance the condition for maximum power output is known as quarter-wave resonance as the transducer has to be an odd integral number of quarter-wavelengths thick. At this resonance the velocity of the front surface of the transducer at $x = l$ is given by:

$$\frac{\partial \xi_l}{\partial t} = \frac{hC_oV}{Z_2} \quad . \quad . \quad . \quad . \quad . \quad (3.19)$$

About the only way to get an infinite impedance is to couple a zero impedance to the back of the transducer via a quarter-wave transformer which can be arranged by backing the quarter-wave transducer by another one to make a transducer a half-wavelength thick. If the impedance at $x = 0$ is neither zero nor infinite the condition for maximum power output is:

$$\tan \frac{\pi l}{\lambda} = \frac{Z_1Z_2 + Z_c^2}{2Z_cX_c} \quad . \quad . \quad . \quad . \quad (3.20)$$

$$\approx \left(1 + \frac{Z_1Z_2}{Z_c^2}\right) 150 \text{ (for quartz)}$$

that is:

$$(2n + 1)\frac{\pi}{2 \cdot 01} \leqslant \frac{\pi l}{\lambda} \leqslant (2n + 1)\frac{\pi}{2}$$

Again it is sufficiently accurate for all practical purposes to write:

$$l = (2n + 1)\frac{\lambda}{2} \quad . \quad . \quad . \quad . \quad . \quad (3.21)$$

where $n = 0, 1, 2, 3, 4 \ldots \ldots \ldots$
This condition that the transducer be an odd integral number of half-wavelengths thick is known as half-wave resonance and is the condition under which transducers are normally operated. The wavelength of course is that of the sound within the transducer itself. When the transducer is mechanically resonant the velocity at $x = l$ is given by:

$$\frac{\partial \xi_l}{\partial t} = \frac{2hC_oV}{Z_1 + Z_2} \quad . \quad . \quad . \quad . \quad (3.22)$$

Thus the resonant power output, W_o, from the transducer into a medium of acoustic impedance Z_2 is given by:

$$W_o = Z_2 \left(\frac{\partial \xi_l}{\partial t}\right)^2 = \frac{4h^2 C_o V^2 Z_2}{(Z_1 + Z_2)^2} \qquad . \quad . \quad (3.23)$$

Maximum power output, at mechanical resonance, occurs when $Z_1 = 0$. This condition can be approximated when the transducer is radiating into a solid or liquid by backing it with air which has a very low impedance compared with that for either a solid or liquid.

3.6 Approximate Equivalent Circuits for Transducers

While the equivalent circuit given in Fig. 3.8 is exact (assuming no coupling to other modes) it is too complicated for reasonably easy analysis of transducer operation at frequencies near mechanical resonance. We therefore require a simpler circuit. The mechanical impedance seen by the transformer secondary is:

$$Z_{\text{mech}} = jX_c - j\frac{Z_c}{2} \cot \frac{\pi l}{\lambda} + \frac{jZ_c(Z_1 + Z_2) \tan \frac{\pi l}{\lambda}}{2\left[Z_1 + Z_2 + 2jZ_c \tan \frac{\pi l}{\lambda}\right]} +$$

$$+ \frac{Z_1 Z_2}{Z_1 + Z_2 + 2jZ_c \tan \frac{\pi l}{\lambda}} \quad (3.24)$$

Near resonance and with a low-impedance backing $Z_1 Z_2 \ll Z_c(Z_1 + Z_2)$ $\tan \frac{\pi l}{\lambda}$ so that the last term on the right-hand side of equation (3.24) may be neglected. Therefore, we may now write equation (3.24) as:

$$Z_{\text{mech}} = jX_c - j\frac{Z_c}{2} \cot \frac{\pi l}{\lambda} + \frac{\left[\frac{Z_1 + Z_2}{4}\right] j\frac{Z_c}{2} \tan \frac{\pi l}{\lambda}}{\frac{Z_1 + Z_2}{4} + j\frac{Z_c}{2} \tan \frac{\pi l}{\lambda}} \quad (3.25)$$

As Z_{mech} is the mechanical impedance across the transformer secondary terminals we may now draw the equivalent circuit as in Fig. 3.9. Unfortunately while this circuit is valid looking from the transformer secondary towards Z_1 and Z_2 it is not valid looking from Z_2 into the transducer since, if the transducer is, say, electrically open circuited the mechanical impedance seen by Z_2 in Fig. 3.8, near resonance, is:

$$Z_1 + jZ_c \tan \frac{2\pi l}{\lambda} \quad . \quad . \quad . \quad . \quad (3.26)$$

While that given by the circuit of Fig. 3.9 is:

$$\frac{Z_1}{4} + j\frac{Z_c}{4} \tan \frac{2\pi l}{\lambda} \quad . \quad . \quad . \quad . \quad . \quad (3.27)$$

The discrepancy between expressions (3.26) and (3.27) which comes about because Z_1 and Z_2 in Fig. 3.9 are both divided by four can be rectified by inserting another transformer of turns ratio 2 : 1 as shown in Fig. 3.10.

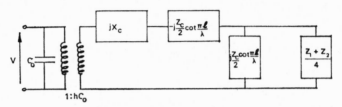

Fig. 3.9 Simplified equivalent circuit for a piezoelectric transducer corresponding to equation (3.25).

The trigonometric impedances in the equivalent circuit of Fig. 3.10 would be more conveniently expressed in normal electrical impedance form. If we consider the series arm impedance $-j\frac{Z_c}{2} \cot \frac{\pi l}{\lambda}$, this goes to zero at the transducer resonance and it would appear likely that it could be represented by a series resonant circuit. We have:

$$-j\frac{Z_c}{2} \cot \frac{\pi l}{\lambda} = -j\frac{Z_c}{2} \tan \left(\frac{\pi}{2} - \frac{\pi l}{\lambda}\right) \rightarrow j\frac{Z_c}{2}\frac{\pi l}{\lambda} - j\frac{Z_c\pi}{4} \left(\frac{l}{\lambda} \rightarrow \frac{1}{2}\right) (3.28)$$

The impedance of a series resonant circuit of inductance L_1 and capacitance C_1 is given by:

$$j\omega L_1 - \frac{j}{\omega C_1} \quad . \quad . \quad . \quad . \quad . \quad (3.29)$$

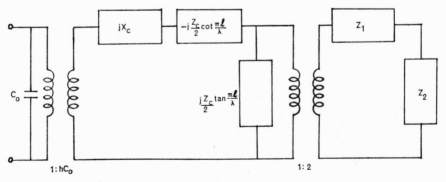

Fig. 3.10 Corrected version of Fig. 3.9 to give correct expressions for both electrical and mechanical impedances.

Hence, equating terms of expressions (3.28) and (3.29) leads to:

$$L_1 = \frac{\rho_c la}{4} \quad \ldots \ldots \quad \textbf{(3.30a)}$$

$$C_1 = \frac{4l}{\pi^2 \rho_c c^2 a} \quad \ldots \ldots \quad \textbf{(3.30b)}$$

For the parallel arm impedance $j\frac{Z_c}{2} \tan\frac{\pi l}{\lambda}$, this goes to infinity at the transducer resonance, indicating that this impedance could be represented by a parallel resonant circuit of inductance L_2 and capacitance C_2, which are easily found to be given by:

$$L_2 = \frac{\rho_c la}{\pi^2} \quad \ldots \ldots \quad \textbf{(3.31a)}$$

$$C_2 = \frac{l}{\rho_c c^2 a} \quad \ldots \ldots \quad \textbf{(3.31b)}$$

Using these values of inductance and capacitance we arrive at the equivalent circuit given in Fig. 3.11.

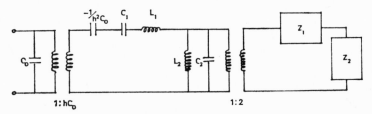

Fig. 3.11 Simplified equivalent circuit for a piezoelectric transducer using standard electrical components and which is valid near to resonance.

There is one further approximate equivalent circuit involving only one mechanical inductance and capacitance and, in practice, is the one most commonly employed since it is the simplest. It is shown in Fig. 3.12

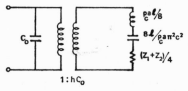

Fig. 3.12 Simplest equivalent circuit for a piezoelectric transducer but which is only valid at frequencies close to resonance.

but its derivation is best left until after we have considered the 'Q' of transducers.

3.7 Electrical Impedance of Piezoelectric Transducer

Since a piezoelectric transducer is driven by an electrical generator, we require to know what impedance it presents and, in particular, how this varies with frequency near the resonant frequency. For this purpose we take the equivalent circuit of Fig. 3.9, transfer all the mechanical impedances through the transformer by multiplying each by $1/h^2C_o^2$ and then calculate the total input impedance, expressing it in the form $R_s + jX_s$, where R_s and X_s denote the resistance and reactance, respectively, of the series circuit which has an impedance equal to that of the transducer. After some manipulation we find, for frequencies close to resonance, that:

$$R_s = \frac{1}{h^2C_o^2} \cdot \frac{\left[\dfrac{Z_1 + Z_2}{4}\right]\left[\dfrac{X_c}{Z_c}\right]^2}{\left[\dfrac{Z_1 + Z_2}{4Z_c}\right]^2 + \frac{1}{4}\cot^2\dfrac{\pi l}{\lambda}} \quad \cdots \quad \text{(3.32a)}$$

$$X_s = -X_c + \frac{X_c^2 \cot \dfrac{\pi l}{\lambda}}{h^2C_o^2 2Z_c\left[\left(\dfrac{Z_1 + Z_2}{4Zc}\right)^2 + \frac{1}{4}\cot^2\dfrac{\pi l}{\lambda}\right]} \quad \text{(3.32b)}$$

where X_c denotes the reactance of C_o. One point worth noting here is that X_c has the same magnitude but opposite sign to the transferred impedance of the negative mechanical capacitance.

Both R_s and X_s can be measured on a bridge and could be plotted as functions of frequency. In practice, however, both quantities are plotted simultaneously on an Argand diagram, as this representation yields useful information more directly than when each quantity is plotted separately. Provided that $(Z_1 + Z_2) \ll Z_c$, equations (3.32) indicate that the graph of X_s plotted against R_s should be a circle which touches the X_s axis and has its centre at the point with co-ordinates $2h^2C_o^2X_c^2/(Z_1 + Z_2)$, $-jX_c$. Although $X_c = 1/\omega C_o$ and is therefore a function of frequency, the frequency range required to give most of the circle is so small that we may regard X_c as being constant. From the position of the circle's centre and the fact that it touches the imaginary axis we note that the circle's diameter is $4h^2C_o^2X_c^2/(Z_1 + Z_2)$, that is the lighter the transducer is loaded the greater is the diameter.

Circle diagrams plotted from experimental data do not touch the imaginary axis as equations (3.32) require but are displaced to the right. This displacement is due to the dielectric loss of the transducer material, a factor which we have so far ignored, and we can obtain an expression for this displacement from the following considerations. At the point of closest approach of the circle to the imaginary axis R_s is a minimum. According to equation (3.32a) this occurs when $\cot(\pi l/\lambda)$ is a maximum, that is, at a frequency equal to twice the transducer resonant frequency (strictly speaking at this frequency equations (3.32) are no longer valid but we may regard this point on the circle as being obtained by extrapolation of the almost complete circle obtained over a narrow frequency range around resonance). At this frequency the transducer will look like a simple parallel combination of capacitance C_o and dielectric loss resistance R_d since the mechanical impedance appearing across the transformer secondary is infinite. The impedance of this connection of C_o and R_d may be written as:

$$\frac{X_c{}^2}{4R_d} - j\frac{X_c}{2}$$

provided that $R_d \gg X_c/2$ where $X_c/2$ is the reactance of C_o at twice the resonant frequency. Thus the series resistance at the minimum value of R_s is $X_c{}^2/4R_d$ and this will be the displacement of the circle. The analysis which we have given is not strictly accurate since while it gives the correct value for the displacement it gives an incorrect value for the reactance.

3.8 Determination of Transducer Efficiency from the Circle Diagram

One important use to which circle diagrams can be put is the determination of transducer efficiencies. We suppose that we have obtained circle diagrams for a transducer when operated under its normal conditions as to loading and also when very lightly loaded, for example, the first might be when radiating into a liquid load and the second when radiating into air. Under these conditions we obtain circle diagrams such as shown in Fig. 3.13. At the resonant frequency (shown as point P) the equivalent circuit is that shown in Fig. 3.12. To a reasonable approximation, Fig. 3.12 can be redrawn as Fig. 3.14 showing that R_d can be incorporated into the circuit as a simple parallel combination with the electrical resistance representing the radiated energy. Thus we have that the diameter of the circle for the loaded transducer is given by:

$$d = \frac{\left[R_d + \dfrac{Z_1 + Z_2}{4h^2 C_o{}^2}\right] X_c{}^2}{R_d \left[\dfrac{Z_1 + Z_2}{4h^2 C_o{}^2}\right]} \quad . \quad . \quad . \quad (3\cdot33)$$

The diameter d' of the unloaded transducer is given by equation (3.33) with $Z_2 = 0$, that is:

$$d' = \frac{\left[R_d + \dfrac{Z_1}{4h^2 C_o{}^2}\right] X_c}{R_d \left[\dfrac{Z_1}{4h^2 C_o{}^2}\right]} \quad . \quad . \quad . \quad (3\cdot34)$$

In this case Z_1 represents the resistive term arising from both the energy radiated into the transducer mounting and the mechanical energy loss

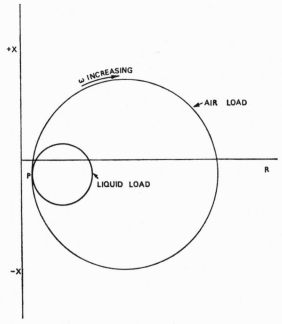

Fig. 3.13 Circle diagrams for lightly and heavily loaded transducers.

within the transducer itself—this is usually small but not always negligible.

As before the displacement d'' is given by:

$$d'' = \frac{X_c^2}{4R_d} \qquad \cdots \cdots \cdots \quad (3\cdot35)$$

The total energy input to the transducer is dissipated in the three resistances R_d, $Z_1/4h^2C_o^2$ and $Z_2/4h^2C_o^2$ but the useful power output appears only in $Z_2/4h^2C_o^2$, the load. Therefore the efficiency η which is defined as:

$$\eta = \frac{\text{power radiated into } Z_2}{\text{total power input}}$$

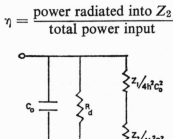

Fig. 3.14 Equivalent circuit for a piezoelectric transducer at resonance.

is given by:

$$\eta = \frac{4h^2 C_o^2 Z_1 R_d}{(Z_1 + Z_2)(Z_1 + Z_2 + 4h^2 C_o^2 R_d)} \quad \cdot \quad \cdot \quad (3\cdot36)$$

Using equations (3.33), (3.34) and (3.35) we have:

$$\eta = \left[\frac{d' - d}{d' - 4d''}\right]\left[\frac{d - 4d''}{d}\right]. \quad \cdot \quad \cdot \quad \cdot \quad (3\cdot37)$$

For transducers with very low electrical losses $d'' \ll d, d'$, so that equation (3.37) can be approximated by:

$$\eta \approx \frac{d' - d}{d'} \quad \cdot \quad \cdot \quad \cdot \quad \cdot \quad \cdot \quad (3\cdot38)$$

Since the diameter of the circle is proportional to X_c^2 it can become very small for a high-frequency transducer particularly when it is heavily loaded and in this case the determination of the efficiency can be difficult. If it is possible to obtain a reasonably sized circle when the transducer is operated with an air load, the efficiency can still be found, provided electrical losses are small, from a modified form of equation (3.38). It can be shown[3.8] that equation (3.38) can be expressed in the following form:

$$\eta \approx \frac{\left(\dfrac{Z_2}{Z_c}\right)\dfrac{1}{k_c^2}}{\left(\dfrac{Z_2}{Z_c}\right)\dfrac{1}{k_c^2} + \dfrac{X_c}{d'}[1\cdot27 + k_c^2]} \quad \cdot \quad \cdot \quad \cdot \quad (3\cdot39)$$

where k_c is the so-called electromechanical coupling coefficient which ranges from 0·1 for quartz to 0·5 for barium titanate. Efficiencies obtainable in practice depend markedly on how the transducer is mounted and can range from about 75% for ceramics up to 99% for quartz.

3.9 Electrical Impedance of Transducer (*cont.*)

In section 3.7 we showed how a transducer can be represented by a series electrical circuit. We can also represent the transducer by a parallel circuit consisting of a resistance R_p and a reactance X_p both in parallel with the static capacitance C_o. Using the approximate equivalent circuit of Fig. 3.9 we eventually obtain, after considerable manipulation of the equations the following expressions for R_p and X_p:

$$R_p = \frac{1}{h^2 C_o^2} \frac{\left[\dfrac{Z_c^2}{4} - \dfrac{X_c Z_c}{2}\tan\dfrac{\pi l}{\lambda}\right]^2 + \left[\dfrac{Z_1 + Z_2}{4}\right]^2\left[X_c - Z_c\cot\dfrac{2\pi l}{\lambda}\right]^2}{\left[\dfrac{Z_1 + Z_2}{4}\right]\left[\dfrac{Z_c^2}{4} - \dfrac{X_c Z_c}{2}\right]}$$

$$+ \frac{[Z_1 + Z_2]Z_c}{8}\left[X_c - Z_c\cot\dfrac{2\pi l}{\lambda}\right]\tan\dfrac{\pi l}{\lambda} \quad (3.40)$$

$$X_p = \frac{1}{h^2 C_o{}^2} \frac{\left[\dfrac{Z_c{}^2}{4} - \dfrac{X_c Z_c}{2} \tan \dfrac{\pi l}{\lambda}\right]^2 + \left[\dfrac{Z_1 + Z_2}{4}\right]^2 \left[X_c - Z_c \cot \dfrac{2\pi l}{\lambda}\right]^2}{\left[\dfrac{Z_1 + Z_2}{4}\right]^2 \left[X_c - Z_c \cot \dfrac{2\pi l}{\lambda}\right]}$$
$$ - \frac{Z_c}{2}\left[\frac{Z_c{}^2}{4} - \frac{X_c Z_c}{2}\right] \tan \frac{\pi l}{\lambda} \quad \textbf{(3.41)}$$

Provided that the transducer is not working very close to resonance, equations (3.40) and (3.41) can be simplified by neglecting terms in X_c to give:

$$R_p \approx \frac{1}{h^2 C_o{}^2}\left[\frac{Z_1 + Z_2}{4}\right]\left[1 + \cot^2 \frac{\pi l}{\lambda} + 2\left(2\frac{Z_c{}^2}{(Z_1 + Z_2)^2} - 1\right)\cot^2 \frac{\pi l}{\lambda}\right]$$
$$\textbf{(3.42)}$$

$$X_p \approx -\frac{Z_c}{h^2 C_o{}^2}\left[\frac{\dfrac{Z_c{}^2}{(Z_1 + Z_2)^2} + \cot^2 \dfrac{2\pi l}{\lambda}}{2\dfrac{Z_c{}^2}{(Z_1 + Z_2)^2}\tan \dfrac{\pi l}{\lambda} + \cot \dfrac{2\pi l}{\lambda}}\right] . \quad \textbf{(3.43)}$$

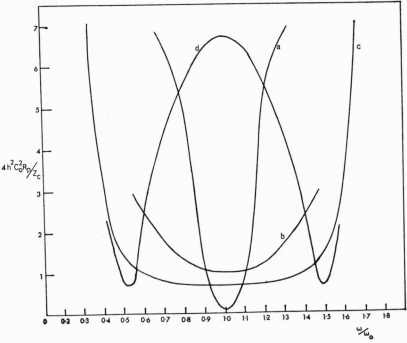

Fig. 3.15 Normalised values for R_p as a function of frequency computed from equation (3.42).

curve	$Z_c/(Z_1 + Z_2)$
a	10
b	1
c	$1/\sqrt{2}$
d	0·15

The forms taken by R_p and X_p are illustrated in Figs. 3.15 and 3.16 which show that they are dependent upon the ratio $Z_c/(Z_1 + Z_2)$. One interesting point is that when $(Z_1 + Z_2) > \sqrt{2}Z_c$ the resistance displays two minima instead of the usual single one.

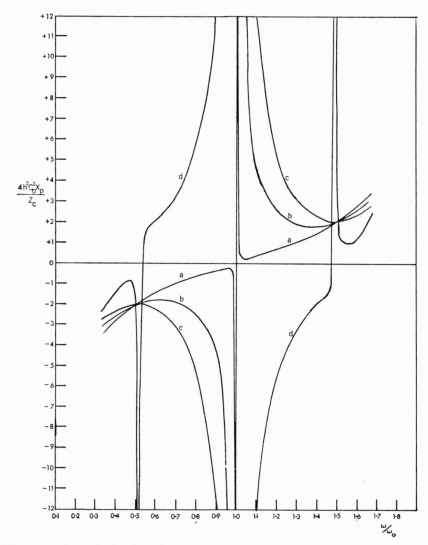

Fig. 3.16 Normalised values for X_p as a function of frequency computed from equation (3.43). Curve designations correspond to those of Fig. 3.15.

Since X_p is in parallel with C_o and is inductive at certain frequencies there is the possibility that X_p and C_o can resonate so that the transducer looks like a pure resistance. This can happen but only if the transducer is very lightly loaded, for example, for quartz $Z_c/(Z_1 + Z_2) > 1\cdot6$ 10^3 for resonance to occur. Usually the reactive impedance X_p is much greater than the reactance of C_o so that the transducer looks like a capacitance C_o shunted by a resistance R_p.

3.10 Mechanical Q of a Transducer

If we assume the transducer to be electrically and mechanically lossless all the input electrical power is converted into acoustic power. For an applied voltage V (r.m.s.) the total radiated power, W_r, is given by:

$$W_r = V^2/R_p \quad . \quad . \quad . \quad . \quad . \quad \text{(3.44)}$$

Since R_p (given by equation 3.40) varies with frequency the radiated power will also vary. In some circumstances, particularly pulsed operation, the transducer has to work over a range of frequencies and so we require some measure of the ability of the transducer to do this. The conventional measure employed in radio engineering is the electrical Q or quality factor of the circuit and, by analogy, we can define a similar quantity Q_{mech}, the mechanical Q, for a transducer. We define Q_{mech} as follows:

$$Q_{\text{mech}} = \frac{f_o}{f_1 - f_2} \quad . \quad . \quad . \quad . \quad . \quad \text{(3.45)}$$

where f_o denotes the mechanical resonance frequency and f_1 and f_2 the frequencies at which the radiated power has fallen to half that at resonance, the applied voltage remaining constant. The condition for the radiated power to have fallen to one half can be obtained from equation (3.42). At resonance the cotangent terms are zero, hence R_p will double, and the radiated power be halved, when the cotangent terms equal unity, that is:

$$\cot^4 \frac{\pi l}{\lambda} + 2\left[2\left(\frac{Z_c}{Z_1 + Z_2}\right)^2 - 1\right] \cot^2 \frac{\pi l}{\lambda} = 1 \quad . \quad \text{(3.46)}$$

In order to be able to use equation (3.45) we assume that f_1 and f_2 are not too far removed from f_o so that we may neglect the term $\cot^4\left(\dfrac{\pi l}{\lambda}\right)$ in equation (3.46). If, furthermore, we assume $2Z_c^2/(Z_1 + Z_2)^2 \gg 1$, which will be the case for liquid loading of the transducer, we have:

$$\cot \frac{\pi l}{\lambda} = \pm\left(\frac{Z_1 + Z_2}{2Z_c}\right) \quad . \quad . \quad . \quad . \quad \text{(3.47)}$$

After some manipulation equations (3.45) and (3.47) give:

$$Q_{\text{mech}} = \frac{\pi}{2}\left(\frac{Z_c}{Z_1 + Z_2}\right) \quad . \quad . \quad . \quad . \quad \text{(3.48)}$$

The importance of equation (3.48) is that, while it is accurate only for lightly loaded transducers, it shows that the greater the frequency range over which the transducer has to work the heavier it should be loaded.

Usually the medium into which the transducer is required to radiate is determined by considerations other than bandwidth but the required bandwidth can be obtained by backing the transducer with a suitable high impedance material. Typical values of Q_{mech} for various loadings and backings are given in table 3.3. These have been calculated from

TABLE 3.3. *Calculated values for the mechanical Q of quartz transducers*

Loading		Backing		
Medium	ρc	Air $\rho c = 420$	Water $\rho c = 1 \cdot 5 \ 10^6$	Loaded epoxy resin* $\rho c = 6 \cdot 5 \ 10^6$
Air	420	$2 \cdot 8 \ 10^4$	15·6	3·6
Water	$1 \cdot 5 \ 10^6$	15·6	7·6	2·9
Quartz	$15 \ 10^6$	1·6	1·4	1·1
Steel	$40 \ 10^6$	0·6	0·6	0·5

Note: Q factors less than 1·2 serve only as a rough guide, since at this value of Q the transducer response becomes double peaked and equation (3.48) becomes inapplicable (cf. section 3.9).

equation (3.48) and include values of Q_{mech} less than 5 although this is really the limit below which equation (3.48) starts to become unreliable although it still acts as a useful guide.

We can use equation (3.48) to derive an equivalent circuit for a piezo-electric transducer, valid near resonance only, which is simpler than those previously given. Near to resonance we can regard the mechanical elements of the transducer as being represented by a series resonant electrical circuit which has the same Q and resonant frequency as the mechanical elements. If we denote the equivalent inductance, capacitance and resistance by L, C and R respectively, we have:

$$Q = \frac{\omega_0 L}{R} = \frac{\pi}{2} \frac{Z_c}{Z_1 + Z_2} \quad \cdots \quad \text{(3.49)}$$

$$\omega_0{}^2 L C = 1 \quad \cdots \quad \text{(3.50)}$$

where $\omega_0 / 2\pi$ denotes the transducer resonant frequency. At resonance the mechanical impedance of the transducer becomes R, hence, from the equivalent circuit given in Fig. 3.10 we have:

$$R = \frac{Z_1 + Z_2}{4} \quad \cdots \quad \text{(3.51)}$$

Thus, combining equations (3.49) and (3.51) gives:

$$L = \frac{Z_c}{16 f_0}$$

*Epoxy resin containing 15% powdered tungsten (cf. reference 8.23)

But since $Z_c = \rho_c c a = \rho_c a f_o \lambda = \rho_c a f_o 2l$
we have:

$$L = \tfrac{1}{8}\rho_c al \qquad . \quad . \quad . \quad . \quad . \quad \textbf{(3.52)}$$

Combining equations (3.50) and (3.52) gives:

$$C = \frac{8l}{\rho_c a \pi^2 c^2}$$

Thus, ignoring the reactance of the negative capacitance, the resulting simple equivalent circuit becomes that shown in Fig. 3.12.

3.11 Electrical Q of a Transducer

The generator used to drive a piezoelectric transducer will normally have a resistive output impedance. Therefore, in order to obtain the maximum transfer of power from the generator to the transducer the latter should present a resistive impedance across its input terminals. As we have seen, a transducer can be represented either by a series or parallel combination of resistance and reactance. This reactance is capacitive and can be neutralised by either a series or parallel inductance which resonates with C_o at the mechanical resonant frequency. Taking the case in which we use a parallel inductance the equivalent electrical circuit for the transducer at resonance is that shown in Fig. 3.17.

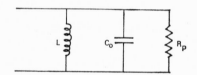

Fig. 3.17 Equivalent circuit for an inductively tuned piezoelectric transducer at resonance.

As far as bandwidth considerations are concerned we now have two resonant circuits, the mechanical one and the electrical one formed by L and C_o. For a wide bandwidth both should have low Q factors. Assuming a lossless transducer and inductance the electrical Q, denoted by Q', of the circuit in Fig. 3.17 is given by:

$$Q' = R_p/\omega_o L$$

For a low mechanical Q, $(Z_1 + Z_2)$, and hence R_p, must be large leading to a large electrical Q. We can counteract this by shunting L with an electrical resistance R_o to bring Q' down to any required value. The effective parallel resistance is now R_p in parallel with R_o which leads to an effective electrical Q, denoted by Q'', given by:

$$Q'' = \frac{Q'}{1 + Q'/\omega_o C_o R_o} \qquad . \quad . \quad . \quad . \quad \textbf{(3.53)}$$

In deriving equation (3.53) we have neglected the electrical Q of the inductance because this is usually high being of the order of 100.

3.12 Transducers as Receivers

Piezoelectric transducers may be used to detect acoustic waves as well as to generate them. If we take the equivalent circuit of Fig. 3.10 and assume, for convenience, that the transducer is electrically unloaded, the voltage V_o appearing across the transducer output terminals for an applied force F is given by:

$$V_o = \frac{F}{hC_o\omega\rho_o Z_c\left(\dfrac{\omega}{\omega_o}\right)\left[\cot^2\left(\dfrac{\pi}{2}\dfrac{\omega}{\omega_o}\right) + \left(\dfrac{Z_1+Z_2}{2Z_c}\right)^2\left(1 - \cot^2\left(\dfrac{\pi}{2}\dfrac{\omega}{\omega_o}\right)\right)\right]^{\frac{1}{2}}}$$

(3.54)

The form taken by this expression is plotted in Fig. 3.18. For frequencies

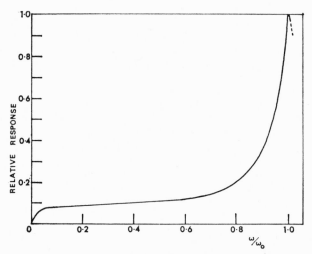

Fig. 3.18 Frequency response of a piezoelectric transducer acting as a receiver.

well below resonance the response is reasonably flat although the output is much smaller than that at resonance. For broadband applications receivers are often operated in this region. However, in many cases the receiver is only required to deal with a narrow range of frequencies so that it can be operated at resonance.

3.13 Resonant Operation of Receiving Transducers

We will assume that the receiving transducer operating at resonance is air backed and shunted electrically by a pure resistance R. If the capacitance C_o is tuned out by means of a suitable inductance the output voltage V_o is given by:

$$V_o = \frac{F}{2hC_o\left[1 + \dfrac{Z_2}{4h^2C_o{}^2R}\right]} \quad . \quad . \quad . \quad (3.55)$$

The form taken by equation (3.55) indicates that we may regard the transducer as a voltage generator with an open circuit output voltage of $F/2hC_o$ and an internal resistance of $Z_2/4h^2C_o{}^2$. Thus for maximum power transfer from the transducer into the load resistance R we have that:

$$R = Z_2/4h^2C_o{}^2 \quad . \quad . \quad . \quad . \quad (3.56)$$

The radiation resistance $Z_2/4h^2C_o{}^2$ can be very large; for example, an air backed 1 mc/s quartz transducer of 1 cm² surface area and with one face in contact with water has a radiation resistance of approximately 1 megohm. A receiving transducer is normally coupled to an amplifier by means of a coaxial cable with a capacitance of the order of 20 pF per foot. At 1 mc/s the reactance of 1 pF is only 160 kilohms so that the cable capacitance must also be neutralised by a shunt inductance, as well as the transducer capacitance C_o, otherwise the signal at the amplifier terminals will be greatly attenuated.

3.14 Sandwich Transducers

If we wish to use a piezoelectric transducer with a large radiating surface or at a low frequency we run into difficulties since in the first case we cannot obtain large enough crystals or manufacture large enough titanate units and in the second case similar considerations apply plus the fact that here the radiation impedance becomes impossibly high so that matching of the transducer to the generator is not possible. Both difficulties may be overcome by the use of a sandwich transducer which consists of a piece, or pieces, of piezoelectric material cemented between two plates of non-piezoelectric material[3.9, 3.10]. By using several pieces of piezoelectric material side by side we can increase the radiating surface. The reduction in transducer resistance comes about because the complete sandwich now forms the resonating system so that the required thickness of piezoelectric material for a given frequency is reduced thereby increasing the turns ratio, hC_o, of the ideal transformer in the transducer equivalent circuit.

The design of sandwich transducers is complicated but we can illustrate one technique as follows. Taking equation (3.24) for the impedance Z appearing across the ideal transformer secondary, neglecting the negative capacitance and assuming that:

$$Z_1 + Z_2 \gg 2Z_c \tan (\pi l/\lambda)$$

we obtain:

$$Z = -jZ_c \cot\left(\frac{\omega l}{\lambda}\right) + \frac{Z_1 Z_2}{Z_1 + Z_2} \quad \cdot \quad \cdot \quad (3.57)$$

If Z_1 and Z_2 are purely resistive the condition for resonance, that is when the imaginary term in equation (3.57) is zero, is that for quarter-wave resonance. If, however, either Z_1 or Z_2 is reactive the resonant frequency differs from that for either quarter- or half-wave resonance. We now assume the backing impedance, Z_1, to be reactive and generated by an acoustic transmission line formed by the rear plate of the sandwich assumed to be terminated in a zero impedance. Thus from equation (1.50) we have:

$$Z_1 = jZ_b \tan\left(\frac{\omega l_b}{c_b}\right) \quad \cdot \quad \cdot \quad \cdot \quad (3.58)$$

where Z_b, l_b and c_b are the characteristic acoustic impedance, thickness and sound velocity respectively for the backing plate. Inserting equation (3.58) into equation (3.57) we find:

$$Z = -jZ_c \cot\left(\frac{\omega l}{c}\right) + j\frac{Z_2^2 Z_b \tan\left(\frac{\omega l_b}{c_b}\right)}{Z_2^2 + Z_b^2 \tan^2\left(\frac{\omega l_b}{c_b}\right)} +$$

$$+ \frac{Z_2 Z_b^2 \tan^2\left(\frac{\omega l_b}{c_b}\right)}{Z_2^2 + Z_b^2 \tan^2\left(\frac{\omega l_b}{c_b}\right)} \quad (3.59)$$

We now assume Z_2 to be resistive, that is the front plate must be either a quarter- or half-wave thick. Assuming the former and denoting the front plate's impedance by Z_f we have:

$$Z_2 = Z_f^2/Z_r$$

where Z_r denotes the impedance of the medium into which the front plate is radiating. If now this medium is a liquid or a gas and if the term $\tan\left(\frac{\omega l_b}{c_b}\right)$ is not too large so that:

$$Z_2^2 \gg Z_b^2 \tan^2\left(\frac{\omega l_b}{c_b}\right)$$

equation (3.39) now becomes:

$$Z = j\left[Z_b \tan\left(\frac{\omega l_b}{c_b}\right) - Z_c \cot\left(\frac{\omega l}{c}\right)\right] + \frac{Z_r Z_b^2}{Z_f^2} \tan^2\left(\frac{\omega l_b}{c_b}\right) \quad (3.60)$$

At resonance the imaginary reactive term must disappear, that is:

$$Z_b \tan\left(\frac{\omega l_b}{c_b}\right) = Z_c \cot\left(\frac{\omega l}{c}\right). \quad \cdot \quad \cdot \quad \cdot \quad (3.61)$$

Thus the electrical resistance of the transducer becomes:

$$R = \frac{1}{h^2 C_o{}^2} \frac{Z_r Z_b{}^2 \tan^2\left(\dfrac{\omega l_b}{c_b}\right)}{Z_f{}^2} = \frac{1}{h^2 C_o{}^2} \frac{Z_r Z_c{}^2 \cot^2\left(\dfrac{\omega l}{c}\right)}{Z_f{}^2}. \quad \textbf{(3.62)}$$

Sanden[3.11] has given details of the design of sandwich transducers. An exact calculation for a brass–quartz–steel system shows that the optimum dimensions are as shown in Fig. 3.19a. Taking $Z_c/Z_f = 0.42$ he finds:

$$R \approx \frac{Z_r}{4h^2 C_o{}^2} 1.3 \quad . \quad . \quad . \quad . \quad . \quad \textbf{(3.63)}$$

For a given radiating surface area the electrical resistance of this transducer is about 4·8 times smaller than that for a simple half-wave quartz transducer operating at the same frequency.

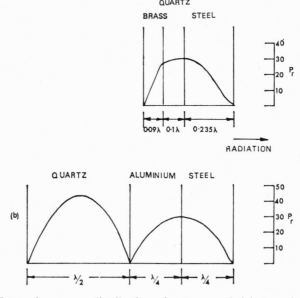

Fig. 3.19 Internal pressure distributions for two sandwich transducers. (After Hueter and Bolt, Sonics, 140.)

An alternative form of sandwich transducer uses two quarter-wave transformers in front of the crystal. For example, in the transducer shown in Fig. 3.19b the front steel plate transforms Z_r to a high impedance which is then transformed by the aluminium plate, which has a lower acoustic impedance than steel, to a value lower than Z_r. This system reduces the transducer resistance by a factor of about seven.

While sandwich transducers are an elegant solution to the problem of either increasing the radiating surface area or reducing the transducer's electrical resistance, practical difficulties can arise since the stresses at the interfaces can be very large and exceed the adhesive strength of cements particularly when the radiated intensity exceeds about 1 watt/cm².

3.15 Variable Frequency Systems

If the layers of a sandwich transducer could be made continuously variable in thickness we would have a continuously tunable transducer. Such a transducer has been built by Fry[3.12] who backed a normal transducer by a variable length mercury column terminated in a low impedance. We can derive the condition for resonance by noting that at resonance the acoustic impedance looking into the transducer must be purely resistive. This impedance consists of the impedance of the mercury column at the rear face of the transducer transformed through the transducer. The column impedance at the rear of the transducer is $jZ_b \tan\left(\dfrac{\omega l_b}{c_b}\right)$ where the suffix b refers to the mercury. After transformation through the transducer the column impedance becomes:

$$Z_c \frac{jZ_b \tan\left(\dfrac{\omega l_b}{c_b}\right) + jZ_c \tan\left(\dfrac{\omega l}{c}\right)}{Z_c - jZ_b \tan\left(\dfrac{\omega l_b}{c_b}\right) . \tan\left(\dfrac{\omega l}{c}\right)}$$

Thus the condition for resonance is:

$$Z_b \tan\left(\frac{\omega l_b}{c_b}\right) + Z_c \tan\left(\frac{\omega l}{c}\right) = 0 . \quad . \quad . \quad (3.64)$$

Fry was able to vary the resonant frequency of his experimental system over a range of about two to one. At the same time the power output varied by about ten to one when the transducer was energised by a constant amplitude voltage.

3.16 Shaped Transducers

So far we have restricted our discussion of transducers to flat plates but since ceramic transducers are made by moulding the ferroelectric material before firing, transducers of any shape can be made and suitably polarised. Normally such shaped transducers are made in the form of bowls or cylinders. A bowl transducer is usually a section of a hollow sphere polarised to execute thickness vibrations so that the radiation from the concave side is focused to pass through a small area at the focal point. Cylindrical transducers are used mainly as receivers since as they can be made very small (outside diameters of 1·6 mm being possible) they interfere only slightly with the acoustic field they are being used to investigate. Like X-cut quartz plates cylindrical receivers have a fairly flat frequency response over frequencies well below the lowest resonant frequency of the cylinder. There are three fundamental resonant frequencies for a cylinder corresponding to length, radial and wall thickness vibrations. There is considerable coupling between the length and radial modes, being strongest when the cylinder's length is between one and two times the diameter. A further complication arises when a cylinder is used as a probe since then it is subject to a uniform external pressure which sets up both radial and tangential stresses. The piezoelectric constants associated with these stresses have opposite signs so that the voltages generated by these stresses tend to cancel. Langevin[3.13] has shown that for good response the ratio of wall thickness to cylinder diameter should lie outside the range of 0·22 to 0·35.

3.17 Transducer Mounting

In practice a piezoelectric transducer must be fitted with electrodes to supply the necessary electrical driving field but these electrodes must not impede the radiation of the ultrasonic waves. It is usual to deposit thin gold films on to the transducer surfaces by means of vacuum evaporation for quartz but for ceramics fired on silver electrodes are used. If the transducer is to radiate into a non-conducting fluid the mounting arrangement shown in Fig. 3.20 is quite satisfactory in which the spring-

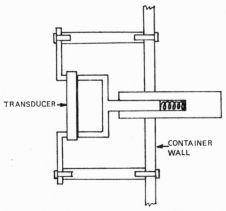

Fig. 3.20 Typical transducer mounting for immersion in a non-conducting liquid.

loaded cup, sealed by a thin metallic diaphragm, presses against the back of the transducer to form an air backing. If a low mechanical Q is required the rear electrode should be replaced by a solid one such as lead- or tungsten-loaded epoxy resin containing a metallic electrode to make electrical contact with the transducer rear surface. In this case the transducer can radiate into a conducting fluid provided that the front electrode is connected to the earth side of the generator and the epoxy resin is carried round the transducer perimeter to form a seal. An alternative technique is to carry the front surface plating around the perimeter of the transducer disc to connect with a ring of plating on the rear surface which is separated from the rear electrode by a narrow unplated anullus. The outer ring of plating on the rear surface can then be sealed to the end of a metal tube by means of a low melting-point alloy such as Wood's metal.

These mounting methods are unsuitable for use at very high frequencies where an initially very thin transducer is being driven at one of its harmonics. At these frequencies an arrangement suggested by Slater[3.14] is used. The transducer consisting of a thin X-cut quartz disc is placed in the gap of a re-entrant cylindrical cavity as shown in Fig. 3.21. In this

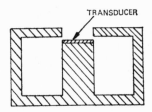

Fig. 3.21 Simplified mounting of a transducer in a resonant cavity.

gap the electrical field is parallel to the crystal X axis so that the disc is set into thickness vibration. With such a system Stewart and Stewart[3.15] detected resonances over a frequency range of from 2950 to 3600 mc/s for a transducer with a fundamental resonant frequency of 27·65 mc/s, implying that the transducer was resonating at harmonics lying between 107 and 129 times its fundamental resonant frequency.

3.18 Very High Frequency Transducers

At very high frequencies the thickness of a transducer vibrating at its resonant frequency becomes extremely small. For example, an X-cut quartz plate would have to be only $2·5\ 10^{-3}$ cm thick if it were to have a half-wave resonant frequency of 100 mc/s. It is obviously impracticable to cut and grind a transducer of this thickness. While it is possible to obtain useful power output from transducers at quite high orders of harmonics, the transducer bandwidth is much more limited than if it were vibrating at half-wave resonance. This is because when vibrating at a harmonic the transducer is several wavelengths thick and therefore a small change in frequency causes a proportionately greater change in, for example, the displacement distribution within the transducer than is the case with a half-wave resonating transducer and it is the relation between the displacement distribution and the transducer boundaries which determines the acoustic output. If we require wideband transducers working at very high frequencies they must, therefore, operate at half-wave resonance. The problem of manufacturing such transducers has only become soluble through the advances in semiconductor technology due to the development of the transistor. Three types of such transducers have appeared so far under the generic name of resistive layer transducers.

3.18a Depletion layer transducers

The depletion layer transducer was the first resistive layer transducer to be developed[3.16]. The electrical resistance of a reverse biased *p–n* junction, formed at the junction of *p*- and *n*-type semiconductors, can be very high whereas the resistance of the individual *p*- and *n*-type semiconductors is small. Hence most of the voltage drop across a device consisting of a slab of *p* material bonded to a slab of *n* material occurs at the junction. This last statement requires qualification at high frequencies since, as the junction looks like a capacitance, we really require that the capacitive reactance of the junction be large compared with the impedances of the bulk of the *p*- and *n*-type slabs. In practice the junction between the two types of semiconductor is formed by suitably doping the pure material with impurities that one part becomes *p* type and the other *n* type. Alternatively a metal semiconductor rectifying junction can be used. The junction sets up a depletion layer in which no current carriers exist; this layer is usually a fraction of a micron thick. If the semiconductor is also piezoelectric, the depletion layer will act like a very thin transducer since the majority of the applied voltage appears across it. Since the width of the depletion layer is proportional to the value of the applied steady reverse voltage we have a voltage tunable transducer for very high frequencies.

3.18b Diffusion layer transducer

Because the depletion layer is so very thin, the depletion layer transducer is only suitable as a half-wave resonator at frequencies exceeding 1000 mc/s. For lower frequencies the diffusion layer transducer[3.17] is more suitable. This type is formed by diffusing a suitable impurity into a plate of piezoelectric semiconductor such that in the region containing the impurity, the impurity traps the current carriers so that the region exhibits a high resistance. This high resistance region acts like the depletion layer in that most of the applied voltage is dropped across it but it has the advantage that its width can be controlled by controlling the diffusion process to give half-wave transducers for frequencies of a few hundred megacycles. Successful diffusion layer transducers have been made by diffusing copper into cadmium sulphide where the copper atoms trap the normally free conduction electrons.

3.17c Epitaxial transducers

An epitaxial transducer can be made by depositing a suitably orientated layer of resistive piezoelectric material on to a conducting substrate. This may be done, for example, by evaporating resistive cadmium sulphide on to conducting cadmium sulphide. Although both materials are the same this is not essential. All that is necessary is that the substrate forms a conducting base of suitable crystalographic properties to give the required orientation to the deposited piezoelectric resistive material.

3.19 The Measurement of Transducer Radiation Patterns

There are two main techniques available for plotting the radiation patterns from transducers. The first involves the point by point measurement of the radiation field using a fine probe while the second gives a qualitative overall picture of the field by means of Schlieren photography.

Any probe used to measure the pressure amplitude at a point in the radiation field must disturb the field by as small an amount as possible. Thus if the acoustic mismatch between the probe and the surrounding fluid is large the probe should be small compared with a wavelength. Various types of probe have been developed with the small cylindrical ceramic type being perhaps the most popular. One such probe, developed at the Pennsylvania State College has an outer diameter of only 1·5 mm; it is fully described in Hueter and Bolt[3.18]. An alternative probe of similar dimensions but which is easier to make has been described by Koppelman[3.19]. This probe makes use of the magnetostrictive effect in nickel. A pure nickel wire is fitted inside a polythene tube sealed at its bottom end as shown in Fig. 3.22. This can be done by melting the bottom of the tube so that a seal is formed and then pushing the wire through this seal so that it protrudes a short distance. The air space between the inside of the polythene tube and the wire ensures that only the exposed end of the nickel wire comes into contact with the ultrasonic radiation. The other end of the wire is loaded with a polythene, rubber or plasticine wedge to absorb the waves travelling up the wire. Just below this wedge is wound a small coil of copper wire which is connected to the input of an amplifier. If the nickel wire is polarised by either passing a direct current through the coil or by placing a permanent magnet near the coil, stress waves travelling up the wire, induced by the ultrasonic pressure variations at the exposed tip, will cause flux changes within the coil and thereby induce a voltage in it. For maximum sensitivity the thickness of the coil should be less than half the wavelength of the stress waves in the nickel wire.

If the fluid in which the ultrasonic waves are travelling is an electrolyte two other types of probe are available. The first is the electrokinetic probe[3.20] which depends upon the fact that if a double cotton covered copper wire is immersed in a sound field in an electrolyte an alternating voltage at the sound frequency is set up between the wire and the solution. The second type, described by Fox, Herzfeld and Rock[3.21], uses two fine wires, insulated except at their tips, placed in an electrolyte with the plane of the wires at right angles to the direction of propagation of the radiation. The ultrasonic pressure variations modulate the electrolyte conductivity between the wire tips. If now an audio frequency current is passed through the wires, the conductivity modulation causes two signals to appear across the wires, one at the sum of the audio and ultrasonic frequencies and one at their difference, whose amplitudes are directly proportional to the amplitudes of the audio current and the ultrasonic pressure. One of these signals is amplified by means of a

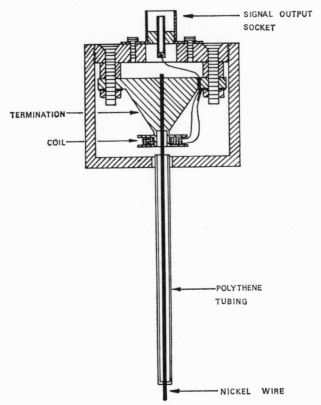

Fig. 3.22 Cross-section through a Koppelman probe.

selective amplifier and displayed on a valve voltmeter. Provided that the relationship between electrolyte conductivity and pressure is known this method provides an absolute measure of pressure amplitude at a point. However, unless the probe leads are carefully screened erroneous signals can be produced by non-linear effects at the probes causing electrical mixing of the picked-up signal at the transducer frequency with the audio current giving rise to signals at the sum and difference frequencies which are undistinguishable from the desired signals.

While all these probes are designed to be as small as possible larger probes can be used provided that their impedances match that of the surrounding medium. One such probe has already been described in section 2.9.

The Schlieren technique[3.22] for the visualisation of the ultrasonic radiation field depends upon the fact that the pressure variations within the medium cause minute variations in the medium's refractive index which are, nevertheless, sufficient to refract light waves. A typical Schlieren apparatus is illustrated in Fig. 3.23. A point source of light is

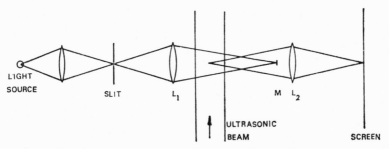

Fig. 3.23 Schematic arrangement for the visualisation of an ultrasonic field using Schlieren techniques.

focused on to a slit whose image is focused on to a mask M by a lens L_1. A further lens L_2 focuses an image of the waves on to a screen. The mask prevents any light from the slit reaching the screen in the absence of the ultrasonic field. As soon as this field is switched on it refracts the light so that some is now no longer prevented by the mask from reaching the screen and upon which appears a picture of the field in which the regions of greater intensity are indicated by brighter illumination.

In connection with the Schlieren technique we might add here that the interaction of ultrasonic and light waves has been suggested as a technique for the modulation and deflection of laser beams.

References

3.1 Lutz, S. G. and Rand, G., *Electronics*, **22**, 12 (1949).
3.2 Hartmann, T., *J. Sci. Inst.*, **16**, 146 (1939).
3.3 Janovsky, W. and Pohlman, R., *Z. Angew. Phys.*, **1**, 222 (1948).
3.4 Allen, C. H. and Rudnick, I. *J. Acoust. Soc. Amer.*, **19**, 857 (1947).
3.5 St. Clair, H. W., *Rev. Sci. Inst.*, **12**, 250 (1941).
3.6 Bordoni, P. G. and Nuovo, M., *Acustica*, **8**, 351 (1958).
3.7 Koehler, G., *Circuits and Networks*, p. 246. Macmillan (1955).
3.8 Hueter, T. F., and Bolt, R. H., *Sonics*, p. 121. Wiley (1955).
3.9 Fry, W. J. and Dunn, F., *J. Acoust. Soc. Amer.*, **34**, 188 (1962).
3.10 Fry, W. J., Taylor, J. M. and Henvis, B. W., *Design of Crystal Vibrating Systems*. Dover (1948).
3.11 von Sanden, K., *Doctorate Thesis, Technische Hochschule*. Hanover (1950).
3.12 Fry, W. J., Fry, R. B. and Hall, W., *J. Acoust. Soc. Amer.*, **23**, 94 (1951).

D

3.13 Langevin, R. A., *J. Acoust. Soc. Amer.*, **26**, 421 (1954).

3.14 Slater, J. C., *J. Acoust. Soc. Amer.*, **29**, 758 (1957).

3.15 Stewart, J. L. and Stewart, E. S., *J. Acoust. Soc. Amer.*, **33**, 538 (1961).

3.16 White, D. L., *I.R.E. Natl. Conv. Record*, Part 6, p. 304 (1961).

3.17 Foster, N. F., *J. App. Phys.*, **34**, 990 (1963).

3.18 Hueter, T. F. and Bolt, R. H., *Sonics*, p. 151. Wiley (1955).

3.19 Koppelman, J., *Acustica*, **2**, 92 (1952).

3.20 Yeager, E., Dietrich, H., Bugosch, J. and Hovorka, F., *J. Acoust. Soc. Amer.*, **23**, 627 (1951).

3.21 Fox, F. E., Herzfeld, K. F. and Rock, G. D. *Phys. Rev.* (2), **70**, 329 (1946).

3.22 Willard, G. W., *Bell Lab. Record*, **25**, 194 (1947).

4
Magnetostrictive Transducers

4.1 Magnetostriction

The name magnetostriction or piezomagnetism is given to the phenomenon that when some materials are magnetised a change in dimensions
occurs. This change can be either positive or negative in a direction
parallel to the magnetic field and is independent of the direction of the
field. If we consider a small element of magnetostrictive material of
length δx situated in a magnetic field of flux density B_o in the x direction
and denote the change in δx due to B_o by ξ, we have:

$$\frac{\xi}{\delta x} = cB_o{}^2 \quad . \quad . \quad . \quad . \quad . \quad (4.1)$$

where c is a constant. For a small change in B_o, which we will denote by
B we have:

$$\frac{\delta \xi}{\delta x} = 2cB_oB \quad . \quad . \quad . \quad . \quad . \quad (4.2)$$

We now suppose that the magnetostrictive material is subject to both
a tensile stress T and a magnetic field so that the resultant strain will be
the sum of the strains due to the mechanical stress and the magnetostrictive effect, that is:

$$\frac{\delta \xi}{\delta x} = \frac{1}{c_x} T + 2cB_oB$$

or:

$$T = c_x\frac{\partial \xi_x}{\partial x} - 2cc_xB_oB . \quad . \quad . \quad . \quad (4.3)$$

where c_x denotes the elastic modulus in the x direction.

Equation (4.3) is analogous to equation (3.4) for piezoelectric materials
with B being analogous to D and $2cc_xB_o$ analogous to h. We would expect to find another magnetostrictive equation analogous to equation
(3.5) with H instead of E and the permeability μ instead of the permittivity ε, that is:

$$H = -2cc_xB_o\frac{\partial \xi}{\partial x} + B/\mu \quad . \quad . \quad . \quad (4.4)$$

If, for convenience, we denote the term $2cc_xB_0$ by λ the two equations for magnetostriction become:

$$T = c_x\frac{\partial \xi}{\partial x} - \lambda B \quad . \quad . \quad . \quad . \quad . \quad (4.5)$$

$$H = -\lambda\frac{\partial \xi}{\partial x} + B/\mu \quad . \quad . \quad . \quad . \quad . \quad (4.6)$$

Since these two basic equations are exactly analogous to those for piezoelectricity we can immediately apply the theory developed in the preceding chapter for piezoelectric transducers to magnetostrictive transducers provided that we take into account the fact that the internal energy losses in a magnetostrictive transducer are more important than the losses in a piezoelectric transducer.

If we neglect the losses of a magnetostrictive transducer, formed by winding a wire coil around a core of magnetostrictive material, its equivalent circuit must be similar to that for a piezoelectric transducer. We therefore require to find the turns ratio of the ideal transformer coupling the electrical to the mechanical network. In the piezoelectric case this turns ratio is given by hC_0, that is, hQ/V. The magnetostrictive analogue of this expression is $\lambda BS/V$. Now assuming that we apply a sinusoidal voltage to a coil wound round the transducer core and that a resulting current i flows we have:

$$V = -L\frac{di}{dt} = -\omega Li \quad . \quad . \quad . \quad . \quad . \quad (4.7)$$

where L denotes the coil inductance. One definition of inductance is that it equals the flux linkages generated by one ampere flowing in the coil, that is:

$$L = BSN/i \quad . \quad . \quad . \quad . \quad . \quad (4.8)$$

where S denotes the cross-section of the core assumed to be constant throughout the core. Inserting equation (4.8) into equation (4.7) gives:

$$BS/V = -(1/\omega N)$$

Hence the turns ratio of the ideal transformer becomes:

$$-\lambda/\omega N$$

If we were to clamp the transducer so that it could not vibrate its ends would each have zero velocity hence no mechanical 'velocity' will flow in the transformer secondary and therefore no electrical current flows in its primary. However, a current still flows through the winding, therefore the winding inductance appears shunted across the transformer primary so that the equivalent circuit of the ideal lossless magnetostrictive transducer is shown in Fig. 4.1. The components of the transmission line are similar to those for a piezoelectric transducer hence the magneto-

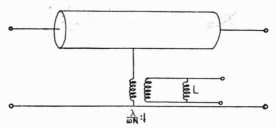

Fig. 4.1 Basic equivalent circuit for a magnetostrictive transducer.

strictive transducer has resonant frequencies given by the same type of expressions, that is the fundamental resonance occurs when the transducer is one half-wavelength long.

4.2 Losses in a Magnetostriction Transducer

The energy losses arising in any magnetic device arise from two sources, the electrical resistance of the winding and the energy dissipated in the core itself.

The resistance of the winding when it carries an alternating current is greater than its value when measured under d.c. conditions because of the skin and proximity effects. The current in a wire sets up its own magnetic field which tends to concentrate the current in the outer layers of the wire so that the effective cross-section of the wire through which the current is flowing is reduced thereby increasing the apparent wire resistance—the so-called skin effect. If two wires, each of which is carrying a current, are placed side by side the magnetic field of one will interact with the current flowing in the other in such a way that its effect is similar to the skin effect thereby further increasing the wire's apparent resistance—the proximity effect.

The core losses arise from two sources, hysteresis and eddy currents. The relation between B and H for a magnetic material is not a single valued function but is as shown in Fig. 4.2. If we apply a sinusoidal

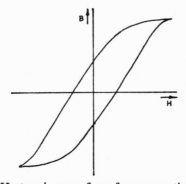

Fig. 4.2 Hysteresis curve for a ferromagnetic material.

voltage across a winding around a core of magnetic material, the flux density B will be in quadrature with the voltage (since voltage is proportional to dB/dt). However, since H lags behind B and since the current into the winding is directly proportional to, and in phase with, H, the current will not be in exact quadrature with the voltage. Therefore some power will be dissipated by the system. The power loss per unit volume due to hysteresis, P_h, is given by the empirical relation:

$$P_h = k_h f B_m{}^n \qquad \qquad \text{(4.9)}$$

where k_h ranges from $6 \cdot 10^{-7}$ to $4 \cdot 10^{-6}$ and n from $1 \cdot 5$ to $2 \cdot 5$ depending upon the material. B_m denotes the maximum value of the flux density.

Eddy current losses arise from the fact that the alternating magnetic flux in the core of the transducer may be regarded as passing through a series of rings of the core material with the planes of the rings perpendicular to the direction of the flux. A voltage, therefore, is induced in each ring which causes a current to flow. This current flows through the resistance of each ring and power is thereby dissipated. We can reduce the power loss by inserting a high resistance into the current path either by using a high resistance magnetostrictive material such as a ferrite or, alternatively, by constructing the core from thin sheets or laminations insulated from each other. Eddy currents still flow within each lamination but their effect is greatly reduced compared with the effect with an unlaminated core. The power loss per unit volume, P_e, is given by:

$$P_e = \frac{\pi^2 T^2}{6\rho} f^2 B^2{}_m \qquad \qquad \text{(4.10)}$$

where ρ denotes the core material resistivity and T the lamination thickness. Usually laminations of between $0 \cdot 2$ and $0 \cdot 05$ cm thickness are used over the frequency range of 10 kc/s to 100 kc/s where magnetostrictive transducers are most commonly used.

For both hysteresis and eddy current losses the power loss is proportional to $B^2{}_m$. Since B_m is directly proportional to the voltage V applied to the winding, the core losses are proportional to V^2, that is they can both be represented by a resistance shunting the winding.

Before we can finalise the equivalent circuit for a practical transducer there is one further imperfection to be taken into account. This is the leakage inductance. Not all the magnetic flux generated by the winding passes through the core, some, as it were, leaks, hence it is termed the leakage flux. This flux, however, still interacts with the winding so that while from the point of view of producing magnetostriction it does not exist nevertheless it exists electrically. We represent it in the equivalent circuit by an inductance, the leakage inductance, placed in series with the winding resistance.

We can now give the complete equivalent circuit which is shown in Fig. 4.3.

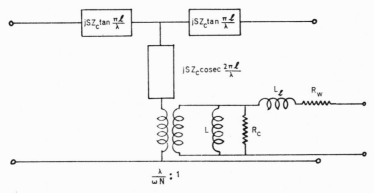

Fig. 4.3 Complete equivalent circuit for a magnetostrictive transducer. L represents the winding inductance, R_w its resistance, L_l the leakage inductance and R_c the core losses. Z_c denotes the core's specific acoustic impedance and S its cross-sectional area.

4.3 Transducer Efficiency

Magnetostrictive transducers find their widest application in the field of ultrasonic processing in which they may be required to generate considerable acoustic power. The efficiency of a transducer is composed of two parts: the efficiency of conversion of electrical into mechanical energy and the conversion of mechanical into acoustic energy. While the latter is relatively constant, at about 80%, over a wide-frequency range, although falling off at the higher frequencies, the former decreases with increasing frequency. This decrease is due to the facts that the hysteresis loss increases linearly with frequency and the eddy current loss increases with the square of the frequency. At low frequencies, around 10 kc/s the electromechanical conversion efficiency is of the order of 90% but has fallen to around 20% at frequencies in the 100 kc/s region. The overall efficiency therefore ranges from typically 70% at around 10 kc/s to about 20% around 200 kc/s.

4.4 Transducer Matching

Maximum electrical power can be fed into a transducer from an electrical generator when the transducer input impedance is purely resistive and equals the output resistance of the generator. The winding inductance can be neutralised by coupling the transducer to the generator through a suitably chosen capacitor so that the capacitor and winding inductance form a series resonant circuit at the transducer's operating frequency. This capacitor also serves to block the polarising current passed through the winding to generate the polarising flux (cf. B_o in equation 4.2). In this case and if the transducer losses are ignored, the input resistance, R, of the transducer is given by:

$$R = \frac{4N^2S^2\lambda^2\mu^2}{l^2\rho_0 c_0 S} \qquad . \quad . \quad . \quad . \quad \textbf{(4.11)}$$

where N denotes the number of turns of the winding, S the radiating surface, μ the effective permeability of the core, l the magnetic path length within the core, and $\rho_0 c_0$ the specific acoustic impedance of the medium into which the transducer is radiating. Equation (4.11) shows that the transducer can be matched to the generator by suitable choice of the number of turns on the winding. In practice, however, it is often more convenient to use a smaller number of turns than equation (4.11) would indicate, in order to reduce the winding resistance losses and at the same time keep the voltage to be applied across the winding as small as possible in order to decrease the insulation problems. Matching is then carried out by inserting a transformer of suitable turns ratio between the transducer and the generator output. This transformer introduces a further source of power loss but with a well-designed unit this loss is usually negligible.

4.5 Transducer Design

The two materials most commonly used for constructing magnetostrictive transducers are pure nickel and iron-cobalt alloys, in particular Permendur composed of 49% iron, 49% cobalt and 2% vanadium. Laminations of these materials are constructed into transducers with the shapes shown in Fig. 4.4. These shapes provide closed paths for the

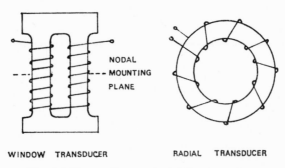

Fig. 4.4 Types of magnetostrictive transducer.

magnetic flux. If a simple laminated rod were used its reluctance would be so high because of the large air path to be traversed by the flux that an unnecessarily high energising current would be required to produce a sufficient magnetic flux to obtain a usable amount of acoustic power output. These transducers are normally operated at their fundamental half-wave resonant frequencies which means that there is a velocity node half-way along the stack. Since there is no motion at this point, the transducer can be rigidly mounted at this point without any energy being lost through the mounting.

An alternative form of mounting, shown in Fig. 4.5, is used when a large amount of acoustic energy is required to be radiated through a small area. A large transducer is coupled by a cone to the point of application of the energy. If this cone is tapered exponentially the vibration amplitude increases inversely with the cone diameter. In this case the transducer and cone must be regarded as a single unit with a velocity node, suitable for a mounting point somewhere along the cone.

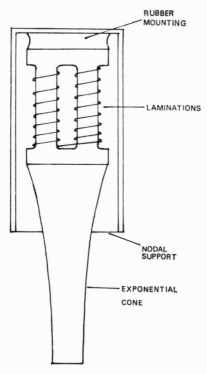

Fig. 4.5 Magnetostrictive transducer with coupling cone.

The detailed design of magnetostrictive transducers is beyond the scope of this book but readers will find useful design data in articles by Notvedt[4.1], Perkins[4.2] and Whymark[4.3].

References

4.1 Notvedt, H., *Acustica*, **4**, 432 (1954).
4.2 Perkins, J. P., *Ultrasonics*, **2**, 193 (1964).
4.3 Whymark, R. R., *Acustica*, **6**, 277 (1956).

5
High Power Ultrasonics

5.1 Introduction

The application of ultrasonics can be divided into two fields; low power and high power. There is no sharp dividing line between these two fields as a particular ultrasonic intensity may in some circumstances be regarded as high power and in others low power. In general, we may regard a particular intensity as being high power if its application results in a permanent change in some property of the irradiated medium or if the non-linear term in the wave equation becomes important. In practice, intensities greater than about 0·5 watt/cm² can be classified as high power. In this chapter we will be concerned solely with high-power ultrasonics.

5.2 Waveform Distortion at High Powers

The solution to the wave equation found in chapter 1 is only valid for waves of infinitesimally small pressure amplitudes. When the pressure amplitude is finite the pressure distribution is no longer sinusoidal but tends towards a sawtooth. That this is so may be seen from the fact that as the velocity of a sound wave is given by $\sqrt{\dfrac{\text{elastic modulus}}{\text{density}}}$ and since the elastic modulus increases faster with pressure than does the density, the velocity of a pressure maximum will be greater than that for a pressure minimum. Thus the peaks of an initially sinusoidal pressure waveform will tend to catch up on the troughs. Although a peak cannot overtake a trough a very steep fronted wave will develop as illustrated in Fig. 5.1. According to Fox and Wallace[5.1] the pressure gradient, at a point x, is related to that at a point x', where $x > x'$, by the relationship:

$$\frac{\partial p}{\partial x} = \frac{\partial p}{\partial x'}\left[1 + \frac{x - x'}{L}\left(\frac{\lambda}{2\pi p}\frac{\partial p}{\partial x'}\right)\right]^{-1} \quad . \quad . \quad . \quad \textbf{(5.1)}$$

where λ denotes the wavelength and L is a characteristic length given by:

$$L = \frac{\rho_0 c_0^2}{\pi}\left(\frac{A}{B}\frac{\lambda}{p}\right). \quad . \quad . \quad . \quad . \quad \textbf{(5.2)}$$

94

where A and B are the coefficient's in the adiabatic equation of state for the medium taken as:

$$p - p_0 = A \frac{\rho - \rho_0}{\rho_0} + \frac{B}{2} \frac{(\rho - \rho_0)^2}{\rho_0{}^2} \quad . \quad . \quad . \quad (5.3)$$

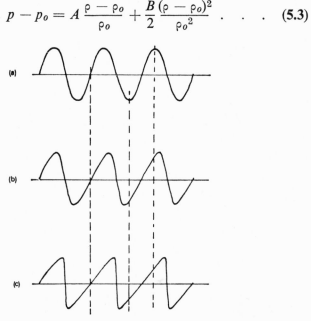

Fig. 5.1 Development of a sawtooth from an initially sinusoidal waveform. Waveforms a, b and c represent the variation in pressure with distance at progressively increasing distances from the generator.

For an initially sinusoidal wave described by the following equation:

$$p - p_0 = P \sin \left(\omega t - \frac{2\pi x'}{\lambda} \right) . \quad . \quad . \quad . \quad (5.4)$$

we have:

$$\frac{\partial p}{\partial x} = - \frac{\dfrac{2\pi P}{\lambda} \cos \left(\omega t - \dfrac{2\pi x'}{\lambda} \right)}{1 - \dfrac{x - x'}{L} \cos \left(\omega t - \dfrac{2\pi x'}{\lambda} \right)} \quad . \quad . \quad (5.5)$$

If we consider the pressure gradients at those parts of the wave where the acoustic pressure is zero, we have, since $\cos (\omega t - 2\pi x'/\lambda) = \pm 1$, that:

$$\frac{\partial p}{\partial x} = \pm \frac{2\pi P/\lambda}{1 \pm \dfrac{x - x'}{L}} . \quad . \quad . \quad . \quad (5.6)$$

The positive sign refers to the trailing edge and the negative sign to the leading edge. Equation (5.6) shows that as x increases the pressure gradient at the leading edge becomes greater while the opposite holds for the trailing edge, that is the originally sinusoidal waveform has degenerated into a triangular wave. When $x - x' = L$ the pressure gradient becomes infinite and a shock wave results. For a gas $A/B = (\gamma + 1)^{-1}$, where γ denotes the ratio of the specific heats (C_p/C_v), so that the distance required for an initially sinusoidal waveform of pressure amplitude P to become a shock wave is given by:

$$L = \frac{\gamma P_o}{\gamma + 1} \frac{\lambda}{\pi P}. \qquad \qquad (5.7)$$

where P_o denotes the static gas pressure.

This shock-wave formation has certain important consequences for the application of high-power ultrasonic radiation in which the desired effects increase with intensity but are independent of the heating effect due to absorption. As the intensity is increased and the waveform becomes more distorted more of the acoustic energy is transferred from the fundamental frequency into its Fourier harmonics. However, the absorption of ultrasonic energy increases as the square of the frequency in liquids and gases so that the distorted wave will show a higher absorption than a pure sine wave[5.2]. In addition, the large pressure gradient at the shock front leads to a large temperature gradient down which heat flows so that this heat energy is lost to the acoustic wave. The resulting increase in absorption on the formation of a shock front implies that in treating a gas increasing the ultrasonic intensity becomes uneconomic since increasing the intensity merely leads to an earlier formation of the shock front and its resultant increased energy loss. In a liquid the same considerations would apply but usually the limiting factor here is cavitation rather than waveform distortion.

5.3 Effects of High Power Radiation

The physical changes induced by intense ultrasonic radiation are caused by one or several of the following: (a) heat, (b) steady ultrasonic forces, (c) cavitation and (d) large mechanical stresses. The large mechanical stresses may be due to cavitation or may arise directly from the large stresses directly associated with an intense ultrasonic wave.

5.4 Heating Effects

When ultrasonic energy travels through a medium it is attenuated to a greater or lesser degree depending upon the medium. The energy abstracted from the radiation appears as heat throughout the volume traversed by the radiation. Thus ultrasonic irradiation provides one technique for the generation of heat within a body and, in particular, if a hemispherical radiator is used intense highly localised heating can be obtained over a small volume deep within a body. This finds its greatest application in medicine for the treatment of muscular disease with moderate intensity levels[5.3] or with higher intensity levels for the localised destruction of tissue[5.4]. Care must be exercised in medical applications that the intensity employed for therapeutic purposes never exceeds the level at which serious permanent damage can occur, unless of course this damage is desired. Even though the intensity is below a danger level, reflection from bone can double the pressure amplitude at a point and therefore quadruple the intensity.

5.5 Steady Ultrasonic Forces

We saw in section 2.8 that the existence of absorption in a fluid gives rise to an energy density gradient which causes streaming. However, even if the fluid possesses negligible absorption but contains suspended particles these particles will experience a steady force since each particle will scatter some of the incident radiation thereby giving rise to an energy density gradient across itself. However, if the suspended particles are much smaller than a wavelength the resulting radiation pressure is small unless the particles are in a standing-wave system. For a rigid sphere of density much greater than that of the surrounding medium and radius r, the force F on the sphere under the influence of radiation of energy density E_o is given by[5.5]:

$$F = 1{\cdot}2(2\pi r/\lambda)^4 E_o \pi r^2 \text{ for progressive waves} \quad . \quad . \quad \textbf{(5.8a)}$$

$$F = 2{\cdot}7(2\pi r/\lambda) E_o \pi r^2 \text{ for standing waves} \quad . \quad . \quad \textbf{(5.8b)}$$

In a standing-wave system the radiation force is directed from regions of high energy density to regions of low density and hence the particles will tend to accumulate in bands situated half a wavelength apart.

As well as radiation pressure, three other unidirectional forces are associated with ultrasonic radiation interacting with suspended particles, namely, the average Stokes pressure, Oseen pressure and Bernoulli attraction.

5.6 Average Stokes Force

If a sphere of radius r is situated in a fluid of viscosity η moving with a velocity u relative to the sphere, it experiences a force F in the direction of the fluid flow given by Stokes' law:

$$F = 6\pi\eta r u \quad . \quad . \quad . \quad . \quad . \quad \textbf{(5.9)}$$

A sphere situated in an acoustic field will experience an alternating force as the fluid oscillates past it under the influence of the acoustic field but in addition it will also experience a steady force. This steady force arises from the fact that the fluid viscosity does not remain constant over one pressure cycle of the acoustic wave because of the temperature variations arising from the adiabatic nature of an acoustic wave. If the fluid is a gas its viscosity increases with temperature so that the viscosity will be greater during the compression part of the cycle than during the rarefaction. Thus the Stokes force acting on the sphere will be greater during compression than during rarefaction and since the direction of the force during compression is opposed to that during rarefaction the sphere will experience a net force.

According to the kinetic theory of gases the viscosity of a gas is given by:

$$\eta = \left(\frac{3kT}{m}\right)^{\frac{1}{2}} \qquad \ldots \ldots \quad \textbf{(5.10)}$$

where k denotes Boltzmann's constant, m the mass of a gas molecule and T the absolute temperature. Also, for adiabatic pressure changes, the absolute temperature T is related to the pressure P by the relation:

$$\frac{T}{T_0} = \left(\frac{P}{P_0}\right)^{\frac{\gamma-1}{\gamma}} \qquad \ldots \ldots \quad \textbf{(5.11)}$$

If now we consider a particle suspended in a sound field of velocity amplitude U_a and pressure amplitude P_a and assume that the particle is sufficiently large so as not to move with the gas molecules, the instantaneous force F acting on the particle is given by:

$$F = 6\pi\eta_0 r \left[\frac{P_0 + P_a \sin \omega t}{P_0}\right]^{\frac{\gamma-1}{2\gamma}} U_a \sin \omega t \qquad . \quad \textbf{(5.12)}$$

where η_0 denotes the gas viscosity at the ambient pressure P_0. Averaging over one complete cycle gives the net force, F_{av}, acting on the particle, that is:

$$F_{av} = \frac{3}{2}\pi\eta_0 r \frac{U_a^2(\gamma-1)}{c} \qquad \ldots \ldots \quad \textbf{(5.13)}$$

where c denotes the sound velocity. In deriving equation (5.13) we have assumed P_a to be much less than P_0.

5.7 Oseen Pressure

The average Stokes force arises from the dependence of the gas viscosity upon temperature leading to a non-linearity which gives rise to a unidirectional force. If the acoustic waveform is non-sinusoidal because the non-linear terms of the wave equation are not negligible a further unidirectional force, known as the Oseen force or pressure, occurs.

We saw in section 5.2 that an initially sinusoidal sine wave tends after a time to become a sawtooth. If we now consider a small particle situated in an acoustic field such that the vibrating fluid molecules transfer some of their motion to the particle the rate of change of momentum of the particle as the front edge of the sawtooth passes it will be greater than during the passage of the less steep trailing edge. Since force equals the rate of change of momentum the force acting on the particle during the passage of the front edge will be greater than during the passage of the trailing edge, that is the particle will experience a net force—the Oseen force. Westervelt[5.6] has shown that this force F can be expressed as:

$$F = 2 \cdot 25\pi r^2 \rho_0 [U \mid U \mid]_{av} \quad . \quad . \quad . \quad (5.14)$$

The time average term $[U \mid U \mid]_{av}$ depends upon the magnitude U^2 and phase θ of the second harmonic component. Denoting the fundamental velocity magnitude by U_0 and letting $b = U_2/U_0$, the net force on the particle is given by:

$$F = - 3r^2 \rho_0 U_0{}^2 b \sin \theta$$

Since the force depends upon $\sin \theta$, both its magnitude and direction will depend upon θ, having a maximum towards the source for $\theta = \pi/2$ and a maximum away from the source for $\theta = -\pi/2$.

For a travelling-wave system the Stokes and maximum Oseen forces are several orders of magnitude greater than radiation pressure. If the wave is very distorted then the maximum Oseen force is greater than the Stokes force. The net force acting on a particle will be the sum of all the steady forces paying due regard to the direction in which each force operates. The radiation force is directed away from the source, Stokes force is directed towards the source while the Oseen force direction depends upon θ.

5.8 Bernoulli Attraction

The motion of a particle suspended in an acoustic field will depend upon its size and we will be discussing the form of this dependence in the next section. However, if the particle is sufficiently large it will not move with the fluid molecules. We consider two such spherical particles of radii r_1 and r_2, with their centres situated a distance d apart on a line at right angles to the direction of propagation of the sound wave. Because of the motion of the fluid molecules under the influence of the sound wave there will be an alternating flow of fluid in the space between the spheres. Applying the equation of continuity to the fluid flow between the spheres, the fluid velocity will be greater than if the spheres were absent. Provided that the flow is laminar this increase in velocity leads to a reduction in pressure δP given by Bernoulli's equation:

$$\delta P = 0 \cdot 5\rho_0(u_1{}^2 - u_0{}^2) \quad . \quad . \quad . \quad (5.15)$$

where u_1 denotes the fluid velocity between the spheres and u_0 the velocity which would exist in the absence of the spheres. Since u_1 will depend upon the sphere geometry and also upon u_0 we may write:

$$\delta P = 0 \cdot 5\rho_0 K u_0{}^2 \quad . \quad . \quad . \quad . \quad (5.16)$$

where K is a geometrical factor. Provided that the sphere diameter is less than a wavelength, the average pressure over one cycle is given by:

$$\delta P_{av} = \tfrac{1}{4}\rho_0 K U_0^2 \quad . \quad . \quad . \quad . \quad . \quad (5.17)$$

where U_0 denotes the fluid velocity amplitude. Since $\rho_0 U_0^2/2 = E_0$, the acoustic energy density, the force of attraction between the spheres, F, will be given by an expression of the form:

$$F = CE_0 \quad . \quad . \quad . \quad . \quad . \quad (5.18)$$

For two spheres C is given by[5.7]:

$$C = \frac{3\pi r_1^3 r_2^3}{d^4} \quad . \quad . \quad . \quad . \quad . \quad (5.19)$$

5.9 Particle Aggregation

The motion taken up by a particle in a sound field, apart from that arising from the steady forces already described, will depend upon the particle's size and mass. Small light particles will move with the fluid whereas large dense ones will not. Intermediate particles will move with an amplitude and phase dependent upon their size and mass.

We consider a spherical particle of density ρ and radius r situated in a medium of viscosity η and a sound field of displacement amplitude ξ_g at a frequency $\omega/2\pi$. If the displacement amplitude of the particle be ξ_p, its equation of motion is found by equating the Stokes force acting on the particle to its inertial reaction to be:

$$6\pi\eta r j\omega(\xi_g - \xi_p) = -\tfrac{4}{3}\pi r^3 \rho\omega^2\xi_p \quad . \quad . \quad (5.20)$$

whence:

$$\xi_p/\xi_g = \cfrac{1}{1 + j\omega r^2\dfrac{2\rho}{9\eta}} \quad . \quad . \quad . \quad . \quad (5.21)$$

and:

$$|\,\xi_p/\xi_g\,| = \cfrac{1}{\left[1 + \left(\omega r^2\dfrac{2}{9\eta}\right)^2\right]^{\frac{1}{2}}} \quad . \quad . \quad (5.22)$$

If we plot $|\,\xi_p/\xi_g\,|$ against r we obtain a curve of the type shown in Fig. 5.2. This indicates that if we irradiate a suspension of different sized particles the larger ones will oscillate with a smaller amplitude than the smaller ones. This difference in oscillation amplitude will cause the particles to have an increased probability of colliding compared to that if they all possessed the same motion. Thus if the particles coalesce on collision we have a means coagulating the suspension.

Coagulation will be the most effective when the velocity difference between particles is a maximum which implies that we should work on the steepest part of the curve in Fig. 5.2, that is in the region where

$| \xi_p/\xi_g | = 0.5$. Substituting this condition into equation (5.22) leads to an expression for the optimum processing frequency of:

$$f_{\text{optimum}} = \frac{9\sqrt{3}}{4\pi} \frac{\eta}{\rho r^2} \quad . \quad . \quad . \quad . \quad (5.23)$$

which for air at N.T.P. is:

$$f_{\text{optimum}} = \frac{22\cdot4\ 10^3}{\rho r^2} \, c/s \quad . \quad . \quad . \quad (5.24)$$

The value for the optimum processing frequency given by equation (5.23) is a guide only since any practical suspension will contain particles of varying sizes distributed about a mean. It is generally better to work

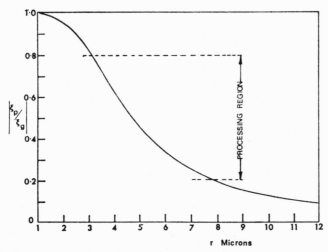

Fig. 5.2 Displacement amplitude for a particle of unity density situated in air relative to that of the air molecules at a frequency of 2 kc/s.

at a lower frequency since the harmonics inevitably present in the processing acoustic field will deal with the smallest particles. As the particles coalesce they will move out of the processing region for the particular frequency employed but aggregation does not necessarily then stop since other mechanisms such as Bernoulli attraction begin to operate.

5.10 Cavitation

If a liquid is irradiated ultrasonically at low power levels no observable effects, apart from slight heating due to absorption, occur. However, if the acoustic pressure amplitude is increased to about 0·25 atmosphere (for water) small bubbles start to appear and which persist after the sound is turned off. On increasing the pressure amplitude to about 1·25 atmospheres small foggy steamers consisting of large numbers of small bubbles, apparently emanating from a point, appear[5.8]. It is at this power level that the physical and chemical effects of ultrasonics become apparent, such as the generation of sound at frequencies which are harmonics and sub-harmonics of the ultrasonic frequency[5.9], the liberation of iodine from potassium iodide solutions[5.10], the degradation of macromolecules[5.11] and the pitting of metallic surfaces[5.12]. These effects do not occur with carefully degassed liquids, until the acoustic pressure amplitude reaches about 4 atmospheres and while the noise generated within the liquid still contains harmonics and sub-harmonics of the ultrasonic frequency an additional continuous white noise signal appears. These three effects are termed, degassing, gaseous cavitation and vaporous cavitation, respectively. Each of these effects can only occur if the liquid contains microscopic weaknesses in the form of minute gas bubbles which can be acted upon by the ultrasonic radiation.

The study of cavitation can be divided into two parts, bubble formation and bubble collapse. We cannot consider bubble formation until we have some idea of the size of bubble involved and as this is obtained from the theory of bubble dynamics in an acoustic field we will consider the growth and collapse of a bubble first, deferring the study of formation until section 5.12.

5.11 Cavity Collapse Theory

The classic theory of the growth and collapse of a cavity in a fluid was developed by Noltingk and Neppiras[5.13] following the early work of Rayleigh and while their theory is necessarily an approximation to the truth in that they neglected the effects of viscosity, it does nevertheless lead to useful results.

We consider a cavity of radius R_0 containing gas at a pressure p_0 situated in a liquid of density ρ under a static pressure P_a. Since the pressure p_0 has to balance both P_a and the pressure due to the liquid's surface tension S, we have:

$$p_0 = P_a + \frac{2S}{R_0} \qquad \cdots \cdots \quad (5.25)$$

If the cavity is situated in a sound field of pressure amplitude P_0 the pressure experienced by the cavity is $P_a + P_0 \sin\omega t$. For the pressure within the cavity to balance that in the surrounding liquid both P_0 and R_0 must vary. The equation of motion for the cavity, which is assumed now to have a radius R at a time t, can be found by noting that the kinetic energy of the liquid must equal the energy expended in expanding the cavity less the work done against the pressure in the liquid and surface tension. Assuming the expansion to be isothermal (which is not strictly accurate), Noltingk and Neppiras give the following equation of motion:

$$2\pi\rho R^3 \left(\frac{dR}{dt}\right)^2 = \int_{R_0}^{R} \left\{ 4\pi R^2 \left[P_a + \left(P_a + \frac{2S}{R_0} \right) \frac{R_0^3}{R^3} - P_0 \sin \omega t \right] - 8\pi RS \right\} dR$$

$$(5.26)$$

The differential of this equation was solved for particular cases by means of a differential analyser and one such solution is reproduced in Fig. 5.3. which shows that provided the initial cavity radius R_o is greater than a critical value the cavity expands and then collapses extremely rapidly. It is this rapid collapse with its associated high pressures which is thought to be responsible for the effects produced by cavitation.

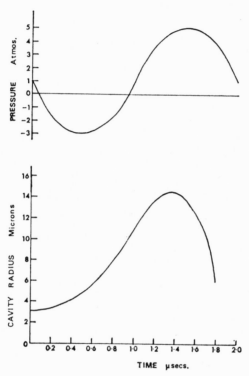

Fig. 5.3 Variation of a cavity radius during one ultrasonic pressure cycle. (After Noltingk and Neppiras (5.13).)

Provided that the time required for the cavity to collapse is much smaller than the period of the ultrasonic wave we may regard the collapse as taking place adiabatically under a constant pressure P equal to the sum of the ambient and peak acoustic pressures. Denoting the maximum cavity radius by R_m, at which the gas pressure within the cavity is Q, and neglecting viscosity, surface tension and liquid compressibility, Noltingk and Neppiras showed that by equating the work done on the system to the kinetic energy of the liquid surrounding the cavity plus the work expended in compressing the gas, the equation of motion of the collapsing cavity in terms of the cavity wall velocity U is:

$$\frac{4\pi}{3}P(R_m{}^3 - R^3) = 2\pi\rho U^2 R^2 + \int_R^{R_m} Q\left(\frac{R_m}{R}\right)^{3\gamma} 4\pi R^2 dR \quad (5.27)$$

where γ denotes the ratio of specific heats for the gas within the cavity. Solving for U gives:

$$U^2 = \frac{2}{3\rho}\left[R(Z-1) - \frac{Q}{1-\gamma}(Z-Z^\gamma)\right] \quad . \quad . \quad (5.28)$$

where $Z = R_m{}^3/R^3$

The pressure within the cavity, q, is given by:

$$q = QZ^\gamma$$

This pressure, q, will be a maximum when Z has a maximum, that is when $U = 0$. Provided that Z is large, the maximum value for Z, Z_{max}, is given by equation (5.28), with $U = 0$, to be:

$$Z_{max} = \frac{P(\gamma-1)}{Q} \quad . \quad . \quad . \quad . \quad (5.29)$$

whence the maximum pressure within the cavity, q_{max}, is given by:

$$q_{max} = Q\left[\frac{P(\gamma-1)}{Q}\right]^{\frac{\gamma}{\gamma-1}} . \quad . \quad . \quad (5.30)$$

The maximum pressure attained by the gas within the cavity can be very large, with correspondingly high temperatures. For example, if $P/Q = 100$, $P = 2$ atmospheres and $\gamma = 1\cdot33$, the peak pressure is 80 000 atmospheres and the peak temperature 10 000° C. Such high temperatures are probably the cause of the faint luminosity[5.14] visible in a cavitating liquid. In practice we would expect the actual peak pressures and temperatures to be less than those calculated from the theory outlined above since the cavity collapse will be slowed down by viscous effects.

In the same paper Noltingk and Neppiras also derived an expression for the pressure distribution in the liquid surrounding the cavity. They found that the pressure p_r at a distance r from the cavity centre is given by:

$$p_r = P - \frac{R}{3r}\left[\frac{QZ^\gamma(3\gamma-4)}{1-\gamma} + \frac{QZ}{1-\gamma} - (Z-4)P\right] - \\ - \frac{R^4}{3r^4}\left[P(Z-1) - \frac{Q(Z-Z^\gamma)}{1-\gamma}\right] \quad (5.31)$$

The form taken by p_r for various values of Z is shown in Fig. 5.4. This pressure waveform will be radiated as a shock wave which, initially at least, has a slow rising leading edge followed by a steep trailing edge which is the reverse of the normal shock front. This waveform has been demonstrated experimentally for the pressure waves generated by spark generated cavities by Guth[5.15] and Mellen[5.16] while the peak liquid pressure close to a cavity has been shown by Ellis[5.17] to be of the same order of magnitude as that predicted by Noltingk and Neppiras.

In order to obtain the most effective use of cavitation we need to know how it is affected by the variables ω, R_o, P_a and P_o. The first point to be noted is that cavitation can only occur if these four variables satisfy certain interrelated conditions[5.18]. We consider a cavity containing a fixed quantity of gas. Since the gas pressure p_o is inversely proportional

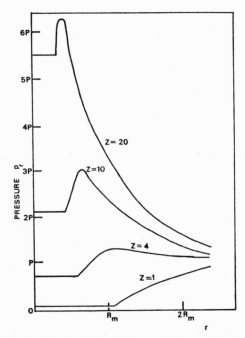

Fig. 5.4 Pressure distribution in the liquid surrounding a collapsing cavity for various values of Z ($= R_m{}^3/R^3$). (After Noltingk and Neppiras (5.13).)

to the cavity volume, we may write $p_o q = K/R^3$,where K is a constant. The equilibrium condition for the cavity is then:

$$\frac{K}{R^3} = P_l + \frac{2S}{R}$$

where P_l denotes the pressure in the bulk of the liquid. If now P_l changes by δP_l the corresponding change in R, δR, is given by:

$$\delta R = - \left[\frac{4S}{R^2} + \frac{3P_l}{R} \right]^{-1} \delta P_l \quad . \quad . \quad . \quad (5.32)$$

Referring to Fig. 5.3 we see that a cavity will only collapse catastrophically if its radius expands as the pressure P_l increases. From equation (5.32) this condition becomes:

$$\frac{4S}{R^2} + \frac{3P_l}{R} < O \quad . \quad . \quad . \quad . \quad (5.33)$$

The most negative value for P_l is $P_a - P_o$, hence a cavity cannot collapse unless its initial radius satisfies the following inequality:

$$R_o > \frac{4S}{3(P_o - P_a)} \quad . \quad . \quad . \quad . \quad \textbf{(5.34)}$$

A cavity cannot collapse completely if its resonant frequency is less than the frequency of the ultrasonic wave since it will not have reached its minimum radius before the ultrasonic pressure will have changed sufficiently to cause it to expand again. The resonant frequency $\omega_r/2\pi$ of a cavity is given by:

$$\omega_r{}^2 = \frac{3\gamma\left(P_a - \dfrac{2S}{R_o}\right)}{\rho R_o{}^2} \quad . \quad . \quad . \quad . \quad \textbf{(5.35)}$$

Exactly what value to take for P_a is arguable but at least equation (5.35) does give a measure of the upper frequency limit for cavitation to occur. Beyond this frequency the cavity does not collapse but oscillates in a complex manner. A further effect of increasing the ultrasonic frequency for values below $\omega_r/2\pi$, is that R_m decreases with increasing frequency.

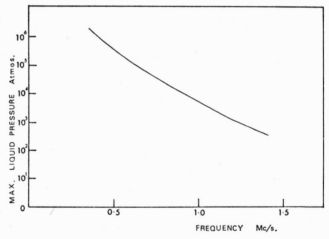

Fig. 5.5 Variation in the maximum pressure in the liquid surrounding a collapsing cavity as a function of the ultrasonic frequency for an ultrasonic pressure amplitude of 4 atmos. and an initial cavity radius of $3.2 \ 10^{-4}$ cm. (After Noltingk and Neppiras (5.13).)

Since the peak pressure in the shock wave from a cavity is proportional to Z^γ, a reduction in R_m reduces the peak pressure. Thus the peak pressure generated decreases with increasing frequency as illustrated in Fig. 5.5.

The maximum cavity radius R_m increases with increasing P_o so that an increase in acoustic intensity should increase the effectiveness of cavitation. This is not necessarily the case since the gas bubbles scatter and absorb the ultrasonic radiation so that an increase in power radiated by a transducer may not necessarily appear as an increased pressure amplitude in the bulk of the irradiated liquid.

The ambient pressure P_a must lie between an upper and a lower limit if cavitation is to occur. As P_a is decreased the cavity resonant frequency decreases so that the cavity collapse time will increase thereby reducing the peak pressure generated. Eventually the cavity resonant frequency will be lower than the ultrasonic frequency and cavitation will cease. As P_a is increased cavitation will again cease once $P_a = P_o$ since then the minimum cavity radius for growth becomes infinite (cf. equation 5.34). Between these two limits for P_a, the peak pressure within the cavity shows a maximum as illustrated in Fig. 5.6.

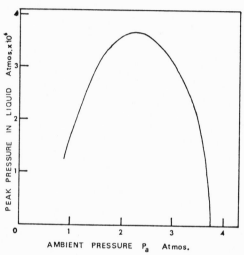

Fig. 5.6 Peak pressure in the liquid surrounding a collapsing cavity as a function of the ambient pressure for an ultrasonic pressure amplitude of 4 atmos. at a frequency of 14·3 kc/s and for an initial cavity radius of 1·6 10⁻⁴ cm. (After Noltingk and Neppiras (5.18).)

5.12 Cavity Formation

So far in discussing cavitation we have assumed the presence of small cavities in a liquid without, however, considering how these are produced. If Noltingk and Neppiras' theory is correct, the initial cavity radius is of the order of 10^{-4} cm implying that the initial gas pressure within the cavity is of the order of 2 atmospheres. This pressure is sufficient to drive the cavity gas into solution. We are thus faced with a dilemma: cavitation exists and requires small cavities yet these cavities cannot, theoretically, exist. There are two possible ways out of this impasse. In the first case we may assume with Fox and Herzfield[5.19] that the cavities are somehow stabilised, either by forming around minute dust particles or that an organic impurity skin of much reduced surface tension exists at the cavity wall. Strasberg[5.20] has attempted to discover the presence of cavities but found that air-saturated water left standing undisturbed for several hours contains less than one bubble per cubic

centimetre with radii in the range of 3 10⁻⁴ to 9 10⁻⁴ cm. In the second
case we may assume with Briggs, Johnson and Mason[5.21] that the cavities themselves are generated ultrasonically. Briggs, Johnson and Mason
start from Eyring's theory of viscosity which assumes the presence of
holes of molecular size in a liquid into which molecules can jump leaving holes behind them. This movement of molecules from hole to hole
gives a liquid its fluidity. In order to jump a molecule must have sufficient energy to surmount an activation potential barrier. At any one
time there will be a few holes of two or three molecules in size. If a
negative pressure of sufficient magnitude is applied to the liquid the
potential barrier for molecules in the direction of a hole is increased
implying that molecules tend to move away from holes thereby enlarg-
ing them. This process is reversed under a positive pressure. However,
if the acoustic pressure exceeds the liquid's cohesive pressure this process is not completely reversible so that the cavities will expand until
they reach a sufficient size to be able to cavitate. We may draw two conclusions from this theory. First, since the rate at which the holes will
coalesce and the cavity grows depends upon the difference between the
peak acoustic pressure and the liquid's cohesive pressure, the acoustic
power must be applied for a certain time before cavitation can occur
and this waiting time will be reduced with an increase in acoustic power.
Secondly, since the acoustic pressure required for this process must be
greater than the cohesive pressure, which is proportional to the hole
activation energy which in turn is proportional to the logarithm of the
liquid's viscosity, the cavitation threshold pressure should be directly
proportional to the logarithm of the viscosity. Briggs, Johnson and
Mason tested these conclusions by applying pulses of acoustic energy
of differing pulse widths and intensities and noting when the onset of

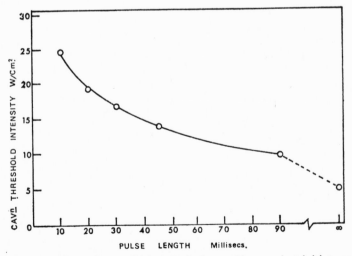

Fig. 5.7 Threshold intensity at which pulsed ultrasonic waves just initiate cavitation
as a function of pulse length. (After Briggs *et al.* (5.21).)

cavitation occurred. Their results, shown in Fig. 5.7 and 5.8, appear to support their theory.

The onset of cavitation in an ultrasonically irradiated liquid is a highly variable phenomenon. Once a liquid has been cavitated it normally cavitates at a lower ultrasonic intensity than that initially required. The cavitation threshold is also lowered if the liquid is irradiated with ionising radiation[5.22] or if electrolysis[5.23] is taking place. Presumably all

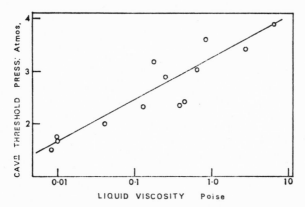

Fig. 5.8 Ultrasonic pressure threshold amplitude as a function of liquid viscosity. (After Briggs *et al.* (5.21).)

these effects are due to the formation of small bubbles within the liquid. Even if only a few are present they will have a large effect since the shock waves which they generate when they cavitate will open up other cavities, that is cavities will breed other cavities.

5.13 Processing with Cavitation

If one reads through the literature dealing with applications of high-power ultrasonic radiation nearly every physical and chemical process seems to have been treated ultrasonically at some time or other. To quote only a few examples, ultrasonically induced cavitation has been found to:

a. produce dispersions of normally indispersable materials such as mercury in nitrobenzene.

b. shatter high molecular weight molecules such as polymers, proteins and viruses.

c. oxidise solutions of potassium iodide to liberate iodine.

d. remove grease and dirt from surfaces.

e. degass liquids.

f. pit the surface of metals.

g. reduce the grain size of electrolytically deposited copper at the surface of an ultrasonically irradiated electrode.

h. increase the rate of chemical reactions.

In some applications ultrasonically induced cavitation may be the only technique currently available for achieving the desired end while in others it may be superior to other techniques. If high-power ultrasonic radiation is to be used on an industrial scale it is important to have some guide as to how the processing may be optimised. The variables under our control are; frequency, intensity, processing tank size and shape, liquid and temperature.

Since the intensity at which cavitation starts increases with frequency the lowest possible ultrasonic frequency should be employed. This frequency, however, should not be below about 20 kc/s since this frequency is approaching the upper frequency of audibility of the human ear. The upper frequency limit should not exceed a few hundred kilocycles/second since above this frequency the cavitation threshold rises rapidly. The ultrasonic intensity need not be much greater than the cavitation threshold since the scattering and absorption of the ultrasonic energy by the cavities tends to cancel any energy increase much above the threshold value.

The tank shape and size will be governed by whether the liquid being irradiated is static or flowing and the volume to be irradiated. If the liquid is to flow a suitable type of arrangement consists of pumping the liquid through a cylindrical transducer vibrating radially. For static systems the transducers are normally mounted at the bottom or along the sides of the tank. If possible a standing-wave system should be set up as thereby greater peak pressures for a given input acoustic power are achieved.

If it is possible to choose the liquid to be irradiated it should be chosen so as to maximise the cavitation intensity. To this end the liquid should have a high surface tension, low viscosity and low vapour pressure. In addition, if the liquid is to be used for ultrasonic cleaning it should wet the objects being cleaned efficiently and also have good solvent properties. At the same time it should not attack the tank chemically and be stable under irradiation. The best liquid for cleaning purposes is perchlorethylene since trichlorethylene, which is normally used for cleaning, tends to be unstable ultrasonically in the presence of water and can form an acidic solution. For straightforward dust removal distilled water to which about 10% of alcohol has been added is satisfactory.

The temperature should be maintained reasonably low in order to keep the liquid's vapour pressure as low as possible.

5.14 Equipment

The efficient transfer of power from the generator into the medium being irradiated is more important at high powers than at low, since apart from economic considerations, any power loss appears as heat which must be dissipated. For maximum power transfer from electrical to acoustic energy the impedances of the transducer and generator must be matched. Ceramic transducers are more suitable than quartz ones since quartz plates of sufficient thickness to make half-wave resonators are practically unobtainable for frequencies in the range of 20 kc/s to 100 kc/s besides requiring very high driving voltages which lead to difficult insulation problems. Sandwich transducers can be used but ceramic plates are more convenient. Ceramic transducers, however, possess appreciable losses so that care must be taken to see that they are not run at such high power levels that their internal power losses cause their temperature to rise above the Curie point. For cavitation processing the transducers are often driven by pulse modulated radio frequency voltages so that the peak power is high enough to cause cavitation while the average power remains low.

If high intensities throughout a small volume are required a focusing radiator is used, either made from a hemispherical solid radiator or else from a mosaic of small elements. In this type of application the effect produced by ultrasonic radiation arises more often from the violent streaming induced in a liquid rather than cavitation and it may be desirable to suppress cavitation altogether by pressurising the system.

The transducer can sometimes be completely immersed in the fluid being irradiated provided that its conductivity is very low. This is not usually the case so the transducer is placed either across an opening in the side of the tank containing the fluid or else placed in good acoustic contact with a side of the tank. If transmission takes place through the tank wall, the wall thickness should either be small compared with a wavelength or else an integral number of half-wavelengths thick. If the transducer is curved the wall of the tank facing the transducer is also curved and given the appropriate thickness. The space between the tank wall and the transducer should be filled with degassed transformer oil, preferably pressurised, to cool the transducer.

The electrical generator to drive the transducer commonly contains valves but transistorised units are becoming available with the advent of high-power high-frequency transistors. The design of such drivers, however, is outside the scope of this book.

References

5.1   Fox, F. E. and Wallace, W. A., *J. Acoust. Soc. Amer.*, **26**, 994 (1954).
5.2   Fox, F. E., *Nuovo Cimento*, 7, Suppl. 2, 198 (1950).
5.3   Tschannen, F. and Sonnenschein, V., *Med. Klinik*, **45**, 1500 (1950).
5.4   Fry, W. J. and Fry, F. J., *I.R.E. Trans. Med. Elec.*, M.E. 7, p. 166 (1960).
5.5   King, L. V., *Proc. Roy. Soc.*, A, **107**, 215 (1947).
5.6   Westervelt, P., *J. Acoust. Soc. Amer.*, **22**, 319 (1950).
5.7   König, W., *Ann. Phys.* (Leipzig), **42**, 43 (1891).
5.8   Willard, G. W., *J. Acoust. Soc. Amer.*, **25**, 669 (1952).
5.9   Esche, R., *Acustica*, 2 Akust. Beih. A.B., 208 (1952).
5.10 Weissler, A. and Cooper, H. W., *Man. Chem.*, **19**, 505 (1948).

5.11 Gooberman, G. and Lamb, J., *J. Poly Sci.*, **42**, 35 (1960).
5.12 Brown, B., *Brit. Commun. Electronics*, **9**, 918 (1962).
5.13 Noltingk, B. E. and Neppiras, E. A., *Proc. Phys. Soc.*, B, **63**, 674 (1950).
5.14 Finch, R. D., *Ultrasonics*, **1**, 87 (1963).
5.15 Guth, W., *Cavitation in Hydrodynamics*, Appendix 1, H.M.S.O. (1956).
5.16 Mellen, R. H., *J. Acoust. Soc. Amer.*, **28**, 447 (1956).
5.17 Ellis, A. T., *Cavitation in Hydrodynamics*, Section 8, H.M.S.O. (1956).
5.18 Neppiras, E. A. and Noltingk, B. E., *Proc. Phys. Soc.*, B, **64**, 1032 (1951).
5.19 Fox, F. E. and Herzfeld, K. F., *J. Acoust. Soc. Amer.*, **26**, 984 (1954).
5.20 Strasberg, M., *Cavitation in Hydrodynamics*, Section 6, H.M.S.O. (1956).
5.21 Briggs, H. B., Johnson, T. B. and Mason W. P., *J. Acoust. Soc. Amer.*, **19**, 664 (1947).
5.22 Sette, D., *Proc. 3rd Int. Cong. Acoust.*, p. 330 (1959).
5.23 Renaud, P. and Saddy, J., *C.R. Acad. Sci. Paris*, **238**, 1393 (1954).

6
Absorption of Ultrasonic Radiation

6.1 Introduction

When an acoustic wave travels through a medium its intensity decreases exponentially with the distance travelled; the energy lost from the radiation appearing as heat. Expressed mathematically we have that the acoustic pressure p_x at a distance x from a point at which the pressure amplitude is P_0, for a wave travelling in the positive direction, is given by:

$$p_x = P_0 e^{-\alpha x} e^{j(\omega t - \beta x)} \qquad \ldots \quad \textbf{(6.1)}$$

where α denotes the 'absorption coefficient' for the medium in which the wave is travelling. As well as possessing absorption all media also show dispersion, that is the wave velocity is frequency dependent. In this chapter we will be considering the causes of absorption and dispersion. For solids the main causes are scattering from individual grains in metals and from dislocations while in fluids absorption can be due to scattering but more important are the 'classical' causes—viscosity and thermal conductivity and the non-classical cause—relaxation mechanisms. This latter phenomenon which arises from the time-lag inherent in any physical or chemical effect is the more interesting since it enables us to use ultrasonic techniques to investigate the molecular structure of matter.

If we differentiate equation (6.1) twice with respect to x and twice with respect to t we obtain:

$$\frac{\partial^2 p_x}{\partial t^2} = \frac{\omega^2}{(\beta - j\alpha)^2} \frac{\partial^2 p_x}{\partial x^2} \cdot \qquad \ldots \quad \textbf{(6.2)}$$

But from equation (1.17) combined with equation (1.13) we have:

$$\frac{\partial^2 p_x}{\partial t^2} = \frac{k + 4\mu/3}{\rho_0} \frac{\partial^2 p_x}{\partial x^2} \cdot \qquad \ldots \quad \textbf{(6.3)}$$

For equations (6.2) and (6.3) to be formally identical we have:

$$\frac{k + 4\mu/3}{\rho_0} = \frac{\omega^2}{(\beta - j\alpha)^2} \cdot \qquad \ldots \quad \textbf{(6.4)}$$

Equation (6.4) shows that, formally, we can regard absorption as being due to k and μ being complex. The problem now is to identify the mechanisms which lead to complex elastic coefficients.

6.2 Classical Absorption due to Viscosity

Equation (6.3) shows that even though we are considering longitudinal wavemotion the shear rigidity μ is involved. A fluid cannot support a static shear but it can support a dynamic one in the form of a viscous drag.

The coefficient of viscosity, η, is defined by:

$$\eta = \sigma_{xy}/(\partial v_x/\partial y)$$

where $v_x = \partial \xi_x/\partial t$
and denotes the fluid velocity in the x direction, hence:

$$\frac{\partial v_x}{\partial y} = \frac{\partial}{\partial y}\frac{\partial \xi_x}{\partial t} = \frac{\partial}{\partial t}\frac{\partial \xi_x}{\partial y} = \frac{\partial e_{xy}}{\partial t}$$

For sinusoidal motion of angular frequency ω the operator $\dfrac{\partial}{\partial t}$ may be replaced by $j\omega$, hence:

$$\sigma_{xy} = j\omega\eta e_{xy}$$

But from equation (1.10) we have:

$$\sigma_{xy} = \mu e_{xy}.$$

hence we have:

$$\mu = j\omega\eta \qquad \cdots \cdots \quad \textbf{(6.5)}$$

If we assume, for the moment that is k not complex, we have from equations (6.4) and (6.5):

$$\frac{\omega^2}{(\beta - j\alpha)^2} = \frac{k + 4j\omega\eta/3}{\rho_0} \qquad \cdots \quad \textbf{(6.6)}$$

Since $\beta = \omega/c$ and $c_0{}^2 = k/\rho_0$, where c_0 denotes the acoustic velocity as ω tends to zero, equation (6.6) can be rewritten as:

$$\frac{1}{(1 - j\alpha/\beta)^2} = \frac{c_0{}^2}{c^2}\left[1 + j\omega\frac{4}{3}\frac{\eta}{\rho_0 c_0{}^2}\right] \qquad \cdots \quad \textbf{(6.7)}$$

The term $\dfrac{4}{3}\dfrac{\eta}{\rho_0 c_0{}^2}$ has the dimensions of time and, for convenience will be represented by τ_L.

After some manipulation, equation (6.7) gives the following equations:

$$\left(\frac{c_0}{c}\right)^2 = \frac{1}{2}\left[\frac{1}{\sqrt{1 + \omega^2\tau_L{}^2}} + \frac{1}{1 + \omega^2\tau_L{}^2}\right] \qquad \cdots \quad \textbf{(6.8)}$$

$$\left(\frac{\alpha c_0}{\omega}\right)^2 = \tfrac{1}{2}\left[\frac{1}{\sqrt{1 + \omega^2\tau_L^2}} - \frac{1}{1 + \omega^2\tau_L^2}\right] \quad . \quad . \quad \textbf{(6.9)}$$

$$\frac{\alpha c_0{}^2}{\omega c} = \tfrac{1}{2}\frac{\omega\tau_L}{1 + \omega^2\tau_L^2} \quad . \quad . \quad . \quad . \quad . \quad . \quad \textbf{(6.10)}$$

If $\omega\tau_L \ll 1$, as is usually the case, $c \approx c_0$, hence equation (6.10) becomes:

$$\alpha = \tfrac{2}{3}\frac{\omega^2\eta}{\rho_0 c_0{}^3} \quad . \quad . \quad . \quad . \quad \textbf{(6.11)}$$

6.3 Absorption due to Thermal Conductivity

Before considering the effect of thermal conductivity upon ultrasonic radiation we must pause to consider the fluid's compressibility K which we find to be more convenient to use than its bulk modulus k. By definition we have:

$$K = -\frac{1}{V}\left(\frac{dV}{dP}\right) \quad . \quad . \quad . \quad . \quad \textbf{(6.12)}$$

Expressing the volume V as a function of temperature T and pressure P we find, eventually:

$$K = K_T - \frac{1}{V}\left(\frac{\partial V}{\partial T}\right)_P\frac{dT}{dP} \quad . \quad . \quad . \quad . \quad \textbf{(6.13)}$$

where K_T denotes the isothermal compressibility. We can obtain a more convenient expression for the second term in the right-hand side of equation (6.13) by a certain amount of manipulation using standard thermodynamic expressions. Eventually we find:

$$K = K_T\left[\frac{\dfrac{dQ}{dT} - \rho_0 V C_v}{\dfrac{dQ}{dT} - \rho_0 V C_p}\right] \quad . \quad . \quad . \quad . \quad \textbf{(6.14)}$$

where Q denotes the quantity of heat added to the system and C_p and C_v denote the specific heats at constant pressure and constant volume respectively. If the process is adiabatic $\frac{dQ}{dT} = 0$, hence:

$$K = K_T/\gamma \quad . \quad . \quad . \quad . \quad . \quad \textbf{(6.15)}$$

During the passage of an ultrasonic wave, the fluid, where there is a high acoustic pressure, will be at a higher temperature than where there is a lower pressure. Heat will, therefore, flow from the higher temperature regions to those with a lower temperature so that the temperature variation along a wave will tend to be smoothed out. This smoothing leads to an energy loss from the wave. The quantity of heat δQ flowing in time δt into a volume V is given by:

$$\delta Q = - K_h V \frac{\partial^2 T}{\partial x^2} \delta t . \quad . \quad . \quad . \quad \textbf{(6.16)}$$

where K_h denotes the coefficient of thermal conductivity of the medium. The temperature T of the medium will follow a similar sinusoidal waveform to the pressure, that is:

$$T = T_a + T_o e^{j[\omega t - (\beta - j\alpha)x]} \quad . \quad . \quad . \quad \textbf{(6.17)}$$

where T_a denotes the ambient temperature and T_o the amplitude of the temperature variations associated with the acoustic wave. Combining equations (6.16) and (6.17) we have:

$$\frac{dQ}{dT} = - K_h V \frac{(\beta - j\alpha)^2}{j\omega} \quad . \quad . \quad . \quad \textbf{(6.18)}$$

If now we combine equation (6.18) with equation (6.4) and assume that the rigidity μ may be neglected, we have:

$$\frac{1}{\rho_0 \omega^2 K_T} (\beta - j\alpha)^2 = \frac{K_h V \dfrac{(\beta - j\alpha)^2}{j\omega} + \rho_0 V C_v}{K_h V \dfrac{(\beta - j\alpha)^2}{j\omega} + \rho_0 V C_p} \quad . \quad \textbf{(6.19)}$$

After some manipulation and under the assumption that $\alpha/\beta \ll 1$ equation (6.19) leads to the following expressions:

$$\alpha = \frac{\omega^2 K_h (\gamma - 1)}{2\rho_0 c^3 C_p} \quad . \quad . \quad . \quad . \quad \textbf{(6.20)}$$

$$\left(\frac{c_o}{c}\right)^2 = 1 + (\gamma - 1)(5 - \gamma)\left(\frac{K_h}{C_p}\right)^2 \frac{\omega^2}{(2\rho_0 c^2)^2} \quad . \quad \textbf{(6.21)}$$

Equations (6.20) and (6.21) are the classical expressions for the absorption and dispersion due to thermal conductivity.

6.4 Relaxation Processes

We saw in section 6.1 that, formally, absorption may be ascribed to the compressibility being complex. Since we are dealing with sinusoidal variations in the thermodynamic parameters describing a medium the appearance of a 'j' term implies that somewhere in the derivation of the expression for compressibility a differentiation with respect to either distance or time has occurred. For example, the derivation of expressions for the absorption due to viscosity involves a time differentiation. For many fluids the measured absorption greatly exceeds that calculated from viscosity and thermal conductivity data so we must look for further time or distance dependent phenomena. In practice we need only consider time dependent effects.

At this juncture it is convenient to consider an electrical circuit analogy. The presence of a 'j' term in an expression relating the voltage and current in an electrical circuit implies that the circuit contains reactance as well as resistance. If instead of applying a sinusoidal voltage to the circuit we apply a step function the current takes a finite time to reach its final steady-state value. Expressed differently, the current relaxes from its initial value to its final value. Expressed in terms of the thermodynamic parameters describing a fluid we have, for example, that if the pressure applied to the fluid is suddenly changed, the volume or temperature may take a finite time to reach is new equilibrium value. This time delay may be due to the time required for the fluid molecules to rearrange themselves. If the temperature is changed suddenly the equilibrium previously existing between the internal and external degrees of freedom of the molecules will be disturbed and a finite time will be required for the new equilibrium to be achieved.

If we denote the time dependent variable by y and its final steady-state value by $\bar{y}$, it is reasonable to assume that the rate at which y approaches $\bar{y}$ will be proportional to $\bar{y} - y$. That is we have:

$$\frac{dy}{dt} = \frac{1}{\tau}(\bar{y} - y) \quad . \quad . \quad . \quad . \quad \textbf{(6.22)}$$

where $1/\tau$ is a proportionality constant. τ has the dimension of time and is termed the relaxation time for the process (cf. τ_L in section 6.2). Processes described by equations similar to equation (6.22) are termed relaxation processes.

6.5 Relaxation Theory

In this section we will obtain an expression for the complex compressibility of a fluid following the method of Andrae and Lamb[6.1]. The expression obtained will be a general one valid for all processes whether structural or thermal. By structural we mean that the energy stored in the internal degrees of freedom changes when a volume change occurs and by thermal we mean that the internal energy changes without any corresponding change in volume. The complete derivation of the complex compressibility is somewhat involved so we will only give the steps of the argument leaving out much of the detailed mathematical manipulation required in order to get the equations into usable forms.

We start by deriving a general thermodynamic expression. When an acoustic wave travels through a fluid it will cause small variations in the pressure P, volume V, temperature T and entropy S at a frequency $\omega/2\pi$. Any three of these four variables will be related at a given frequency by an expression of the form:

$$X = F_\omega(Y, Z) \quad . \quad . \quad . \quad . \quad \textbf{(6.23)}$$

E

where X, Y and Z denote the variables being considered. For small variations in the variables we have:

$$\delta X = \left(\frac{\partial X}{\partial Y}\right)_{Z,\,\omega} \delta Y + \left(\frac{\partial X}{\partial Z}\right)_{Y,\,\omega} \delta Z. \quad . \quad . \quad \textbf{(6.24)}$$

where the partial derivatives have meaning at only the frequency $\omega/2\pi$. We now assume that the frequency dependence of equation (6.23) is due solely to the effects of the internal degrees of freedom of the fluid molecules. Denoting the parameter associated with each internal degree of freedom by y_m we postulate an equation of state of the form:

$$X = F(Y, Z, y_1, y_2, y_3, \ldots \ldots y_m, \ldots \ldots) \quad . \quad \textbf{(6.25)}$$

Differentiating, as for equation (6.24), we obtain:

$$\delta X = \left(\frac{\partial X}{\partial Y}\right)_{Z,\,y} + \left(\frac{\partial X}{\partial Z}\right)_{Y,\,y} + \sum_m \left(\frac{\partial X}{\partial y_m}\right)_{Z,\,y,\,m} \delta y_m . \quad \textbf{(6.26)}$$

where the subscript y indicates that all the y's are held constant during differentiation and the subscript m that all the y's, except y_m, are held constant. Combining equations (6.24) and (6.26) and applying the conditions that Y and ω are to be constant gives:

$$\left(\frac{\partial X}{\partial Z}\right)_{Y,\,\omega} = \left(\frac{\partial X}{\partial Z}\right)_{Y,\,y} + \sum_m \left(\frac{\partial X}{\partial y_m}\right)_{Y,\,Z,\,m} \left(\frac{\partial y_m}{\partial Z}\right)_{Y,\,\omega} . \quad \textbf{(6.27)}$$

The compressibility, K_ω, for an isentropic process at a frequency $\omega/2\pi$ is given by:

$$K_\omega = -\frac{1}{V}\left(\frac{\partial V}{\partial P}\right)_{S,\,\omega}. \quad . \quad . \quad . \quad \textbf{(6.28)}$$

We now expand the right-hand side of equation (6.28) using equation (6.27) in which we substitute V for X, P for Z and S for Y to give:

$$K_\omega = -\frac{1}{V}\left(\frac{\partial V}{\partial P}\right)_{S,\,y} - \sum_m \frac{1}{V}\left(\frac{\partial V}{\partial y_m}\right)_{S,\,P,\,m}\left(\frac{\partial y_m}{\partial P}\right)_{S,\,\omega} \quad \textbf{(6.29)}$$

which may be written as:

$$K_\omega = K_y + \sum_m K_m \quad . \quad . \quad . \quad . \quad \textbf{(6.30)}$$

Since we assume that absorption is due only to the effects of the internal degrees of freedom, K_y, which is the component of the compressibility evaluated for all the internal degrees of freedom held constant, plays no part in causing absorption. We need, therefore, consider only those terms involving K_m. As expressed in equation (6.29) these terms are not directly of use and so we need to transform them into more convenient forms. Writing

$$V = F(P, S, y_1, y_2, \ldots \ldots y_m, \ldots \ldots)$$

we have:

$$\left(\frac{\partial V}{\partial y_m}\right)_{P, S, m} = \left(\frac{\partial V}{\partial y_m}\right)_{P, T, m} - \left(\frac{\partial V}{\partial S}\right)_{P, y}\left(\frac{\partial S}{\partial y_m}\right)_{P, T, m} \quad \textbf{(6.31)}$$

Similarly:

$$\left(\frac{\partial y_m}{\partial P}\right)_{S, \omega} = \left(\frac{\partial y_m}{\partial P}\right)_{T, \omega} + \left(\frac{\partial y_m}{\partial T}\right)_{P, \omega}\left(\frac{\partial T}{\partial P}\right)_{S, \omega}. \quad . \quad \textbf{(6.32)}$$

We now take equation (6.31) term by term.

1. $\left(\dfrac{\partial V}{\partial y_m}\right)_{P, T, m} = \Delta V_m$ —the mth partial volume.

2. $\left(\dfrac{\partial V}{\partial S}\right)_{P, y} = \left(\dfrac{\partial V}{\partial T}\right)_{P, y}\left(\dfrac{\partial T}{\partial S}\right)_{P, y}$

In order to evaluate $\left(\dfrac{\partial V}{\partial T}\right)_{P, y}$ we expand the coefficient of thermal expansion θ_ω by means of equation (6.27) to give:

$$\theta_\omega = \frac{1}{V}\left(\frac{\partial V}{\partial T}\right)_{P, \omega} = \frac{1}{V}\left(\frac{\partial V}{\partial T}\right)_{P, y} + \sum_m \frac{1}{V}\left(\frac{\partial V}{\partial y_m}\right)_{P, T, m}\left(\frac{\partial y_m}{\partial T}\right)_{P, \omega}$$

$$= \theta_y + \sum_m \theta_m \quad \textbf{(6.33)}$$

Hence:

$$\left(\frac{\partial V}{\partial T}\right)_{P, y} = V\theta_y \quad . \quad . \quad . \quad . \quad \textbf{(6.34)}$$

Similarly we expand the specific heat at constant pressure $C_{P, \omega}$ by means of equation (6.27) in order to find $\left(\dfrac{\partial T}{\partial S}\right)_{P, y}$. We have:

$$C_{P, \omega} = T\left(\frac{\partial S}{\partial T}\right)_{P, \omega} = T\left(\frac{\partial S}{\partial T}\right)_{P, y} + \sum_m T\left(\frac{\partial S}{\partial y_m}\right)_{P, T, m}\left(\frac{\partial y_m}{\partial T}\right)_{P, \omega}$$

$$= C_{P, y} + \sum_m C_{P, m} \quad \textbf{(6.35)}$$

Hence:

$$\left(\frac{\partial T}{\partial S}\right)_{P, y} = T/C_{P, y}. \quad . \quad . \quad . \quad \textbf{(6.36)}$$

Combining equations (6.34) and (6.36) we obtain:

$$\left(\frac{\partial V}{\partial S}\right)_{P, y} = \frac{VT\theta_y}{C_{P, y}} \quad . \quad . \quad . \quad . \quad \textbf{(6.37)}$$

3. $\left(\dfrac{\partial S}{\partial y_m}\right)_{P,\,T,\,m} = \dfrac{\Delta H_m}{T}$

where ΔH_m denotes the mth partial enthalpy.

Thus we may now write:

$$\left(\frac{\partial V}{\partial y_m}\right)_{P,\,S,\,m} = \Delta V_m - \frac{V\theta_y\,\Delta H_m}{C_{P,\,y}} = \frac{V\theta_y\,\Delta H_m}{C_{P,\,y}}\left[1 - \frac{\Delta V_m C_{P,\,y}}{V\theta_y\,\Delta H_m}\right] \quad (6.38)$$

We now turn to equation (6.32) which we again take term by term.

1. $\left(\dfrac{\partial y_m}{\partial P}\right)_{T,\,\omega}$. Equation (6.35) gives one expansion for θ_ω. We can find

an alternative expansion since $\left(\dfrac{\partial V}{\partial T}\right)_{P,\,\omega} = -\left(\dfrac{\partial S}{\partial P}\right)_{T,\,\omega}$. Hence we may

write:

$$\theta_\omega = -\frac{1}{V}\left(\frac{\partial S}{\partial P}\right)_{T,\,y} - \sum_m \frac{1}{V}\left(\frac{\partial S}{\partial y_m}\right)_{T,\,P,\,m}\left(\frac{\partial y_m}{\partial P}\right)_{T,\,\omega}. \quad (6.39)$$

Comparing terms of equations (6.33) and (6.39), we find:

$$\left(\frac{\partial S}{\partial y_m}\right)_{T,\,P,\,m}\left(\frac{\partial y_m}{\partial P}\right)_{T,\,\omega} = \left(\frac{\partial V}{\partial y_m}\right)_{P,\,T,\,m}\left(\frac{\partial y_m}{\partial T}\right)_{P,\,\omega}$$

whence:

$$\left(\frac{\partial y_m}{\partial P}\right)_{T,\,\omega} = \frac{\Delta V_m T}{(\Delta H_m)^2}\,C_{P,\,m} \quad . \quad . \quad . \quad (6.40)$$

2. From equation (6.35) and the definition of ΔH_m we obtain:

$$\left(\frac{\partial y_m}{\partial T}\right)_{P,\,\omega} = \frac{C_{P,\,m}}{\Delta H_m} \quad . \quad . \quad . \quad . \quad . \quad (6.41)$$

3. It is a standard thermodynamic identity[6.2] that:

$$\left(\frac{\partial T}{\partial P}\right)_{S,\,\omega} = \left(\frac{\partial V}{\partial S}\right)_{P,\,\omega}$$

Combining this identity with equation (6.37) gives:

$$\left(\frac{\partial T}{\partial P}\right)_{S,\,\omega} = \frac{VT\theta_\omega}{C_{P,\,\omega}} \quad . \quad . \quad . \quad . \quad (6.42)$$

We are now able to obtain a useful expression for $\left(\dfrac{\partial y_m}{\partial P}\right)_{S,\,\omega}$, viz.:

$$\left(\frac{\partial y_m}{\partial P}\right)_{S,\,\omega} = \frac{\theta_y VTC_{P,\,m}}{\Delta H_m C_{P,\,\omega}}\left[1 - \frac{C_{P,\,y}\,\Delta V_m}{V\theta_y\Delta H_m} + \sum_n \frac{C_{P,\,n}}{V\theta_y}\left(\frac{\theta_n V}{C_{P,\,n}} - \frac{\Delta V_m}{\Delta H_m}\right)\right]$$

$$(6.43)$$

(note the summation over the floating index n)

We may refine equation (6.43) slightly by noting that:

$$\frac{V\theta_n}{C_{P,\,n}} = \frac{\left(\dfrac{\partial V}{\partial y_n}\right)_{P,\,T,\,n}\left(\dfrac{\partial y_n}{\partial T}\right)_{P,\,\omega}}{\Delta H_n\left(\dfrac{\partial y_n}{\partial T}\right)_{P,\,\omega}} = \frac{\Delta V_n}{\Delta H_n} \quad . \quad . \quad \textbf{(6.44)}$$

Hence we have:

$$\left(\frac{\partial y_m}{\partial P}\right)_{S,\,\omega} = \frac{\theta_y VTC_{P,\,m}}{\Delta H_m C_{P,\,\omega}}\left[1 - \frac{C_{P,\,y}}{V\theta_y}\frac{\Delta V_m}{\Delta H_m} + \sum_n \frac{C_{P,\,n}}{V\theta_y}\left(\frac{\Delta V_n}{\Delta H_n} - \frac{\Delta V_m}{\Delta H_m}\right)\right]$$

$$\textbf{(6.45)}$$

Combining equations (6.38) and (6.45) we obtain:

$$K_m = \frac{V\theta_y{}^2 TC_{P,\,m}}{C_{P,\,y}C_{P,\,\omega}}\left[1 - \frac{\Delta V_m}{\Delta H_m}\frac{C_{P,\,y}}{V\theta_y}\right]$$

$$\left[1 - \frac{C_{P,\,y}}{V\theta_y}\frac{\Delta V_m}{\Delta H_m} + \sum_n \frac{C_{P,\,n}}{V\theta_y}\left(\frac{\Delta V_n}{\Delta H_n} - \frac{\Delta V_m}{\Delta H_m}\right)\right] \quad \textbf{(6.46)}$$

We can improve equation (6.46) by finding an alternative expression for $\Delta V_m/\Delta H_m$. From the first law of thermodynamics we have:

$$\Delta H_m = \Delta E_m + P_y{}^*\Delta V_m$$

where $P_y{}^* = T\left(\dfrac{\partial P}{\partial T}\right)_{V,\,y}$ denotes the internal pressure. From standard thermodynamic identities we find:

$$\left(\frac{\partial P}{\partial T}\right)_{V,\,y} = \frac{C_{P,\,y} - C_{V,\,y}}{VT\theta_y}$$

whence:

$$\frac{\Delta V_m}{\Delta H_m} = \left[\frac{1 - \dfrac{\Delta E_m}{\Delta H_m}}{C_{P,\,y} - C_{V,\,y}}\right]V\theta_y \quad . \quad . \quad . \quad \textbf{(6.47)}$$

The first term on the right-hand side of equation (6.46) can also be refined by writing some of its components in derivative form to give:

$$\frac{V\theta_y{}^2 TC_{P,\,m}}{C_{P,\,y}C_{P,\,\omega}} = -\frac{1}{V}\left(\frac{\partial V}{\partial P}\right)_{S,\,y}\frac{C_{P,\,m}}{C_{P,\,\omega}}(\gamma_y - 1) = K_y\frac{C_{P,\,m}}{C_{P,\,\omega}}(\gamma_y - 1)$$

$$\textbf{(6.48)}$$

where $\gamma_y = C_{P,\,y}/C_{V,\,y}$

If we now substitute equations (6.47) and (6.48) into equation (6.56), substitute the resulting expression for K_m into equation (6.30) and manipulate the summation signs slightly we obtain:

$$\frac{K_\omega}{K_y} = 1 + \sum_m \frac{C_{P,m}}{C_{P,\omega}} \frac{1}{(\gamma_y - 1)}$$

$$\left[\left(\gamma_y \frac{\Delta E_m}{\Delta H_m} - 1\right)^2 + \frac{\gamma_y^2}{2C_{P,y}} \sum_n C_{P,n} \left(\frac{\Delta E_m}{\Delta H_m} - \frac{\Delta E_n}{\Delta H_n}\right)^2\right] \quad \textbf{(6.49)}$$

Since we have assumed that absorption is due, formally at least, to a complex or time dependent compressibility, only those terms in equation (6.49) which involve a time dependence will contribute to absorption. From the defining equations for ΔV_m, ΔH_m and $P_y{}^*$ and the first law of thermodynamics we can show that:

$$\frac{\Delta E_m}{\Delta H_m} = 1 - \left(\frac{\partial P}{\partial T}\right)_{V,y} \left(\frac{\partial V}{\partial S}\right)_{P,T,m} \quad . \quad . \quad \textbf{(6.50)}$$

Equation (6.50) does not involve any differentiation with respect to any of the y's and hence $\Delta E_m/\Delta H_m$ plays no role in causing absorption. We are left, therefore, with the specific heats.

If we now return to equation (6.22) and assume a sinusoidal perturbation in $\bar{y}$ at a frequency of $\omega/2\pi$ we can replace d/dt by $j\omega$ to give:

$$j\omega y = \frac{1}{\tau}(\bar{y} - y)$$

or, for small changes we have:

$$\delta y = \frac{\delta \bar{y}}{1 + j\omega\tau} \quad . \quad . \quad . \quad . \quad \textbf{(6.51)}$$

Now:

$$C_{P,m} = \Delta H_m \left(\frac{\partial y_m}{\partial T}\right)_P$$

whence:

$$C_{P,m} = \Delta H_m \left(\frac{\partial \bar{y}_m}{\partial T}\right)_P \frac{1}{1 + j\omega\tau_m} = \frac{C_{i,m}}{1 + j\omega\tau_m} \quad . \quad \textbf{(6.52)}$$

where:

$$C_{i,m} = \Delta H_m \left(\frac{\partial \bar{y}_m}{\partial T}\right)_P$$

$C_{i,m}$ represents the steady state or zero frequency value of $C_{P,m}$. If we now substitute equation (6.52) into equation (6.49) we obtain:

$$\frac{K_\omega}{K_y} = \frac{C_{P,y} + \left[\sum_m \frac{C_{i,m}}{1 + \omega^2\tau_m^2}\left\{1 + R_m + \sum_n \frac{R_{m,n}C_{i,n}(1 - \omega^2\tau_m\tau_n)}{1 + \omega^2\tau_n^2}\right\}\right] - j\left[\sum_m \frac{\omega\tau_m C_{i,m}}{1 + \omega^2\tau_m^2}\left\{1 + R_m + 2\sum_n \frac{R_{m,n}C_{i,n}}{1 + \omega^2\tau_n^2}\right\}\right]}{C_{P,y} + \sum_m \frac{C_{i,m}}{1 + \omega^2\tau_m^2} - j\sum_m \frac{\omega\tau_m C_{i,m}}{1 + \omega^2\tau_m^2}} \quad \textbf{(6.53)}$$

where:

$$R_m = \left(\gamma_y \frac{\Delta E_m}{\Delta H_m} - 1\right)^2 \Big/ (\gamma_y - 1)$$

and:

$$R_{m,\,n} = \frac{\gamma_y}{2(C_{P,\,y} - C_{V,\,y})}\left[\frac{\Delta E_m}{\Delta H_m} - \frac{\Delta E_n}{\Delta H_n}\right]^2$$

6.6 Derivation of Absorption and Velocity from Complex Compressibility

If we neglect the effects of viscosity, equation (6.6) may be rewritten in the following form:

$$\left(\frac{c_\infty}{c}\right)^2 - \left(\frac{\alpha c_\infty}{\omega}\right)^2 - 2j\frac{\alpha c_\infty}{\omega c} = \frac{K_\omega}{K_y} \quad \cdots \quad (6.54)$$

where c_∞ denotes the velocity at a frequency sufficiently high that the contribution to the compressibility of all the relaxing components of the specific heat are negligibly small (cf. equation 6.52). Even when only one relaxing mechanism is present the exact solution of equation (6.54) is greatly complicated by the middle term of the left-hand side of the equation, viz. $\left(\frac{\alpha c_\infty}{\omega}\right)^2$. Fortunately this term is small compared with the other real term, $\left(\frac{c_\infty}{c}\right)^2$, so that it may be neglected, particularly in view of the currently available experimental accuracies in the determination of the absorption coefficient.

In order to illustrate the main features of absorption and dispersion we will consider a system in which only one relaxation mechanism exists over the frequency range to be considered. With this condition equation (6.53) reduces to:

$$\frac{K_\omega}{K_y} = 1 + \frac{R_m C_{i,\,m}}{C_{P,\,y} + C_{i,\,m} + j\omega\tau_m C_{i,\,m}} \quad \cdots \quad (6.55)$$

At this stage it is convenient to modify our terminology and denote $C_{i,\,m}$ by C' and $C_{P,\,y} + C_{i,\,m}$ by C_p—the specific heat at constant pressure and zero frequency. We now substitute equation (6.55) into equation (6.54), neglecting the term $\left(\frac{\alpha c_\infty}{\omega}\right)^2$, equate real and imaginary parts and manipulate the resulting equations to obtain the following set of expressions:

$$\alpha\lambda = \frac{\pi R_m C'}{\sqrt{R_m C' C_p + C_p{}^2}} \frac{\omega\tau'''}{(1 + \omega^2\tau'''^2)} \quad \cdots \quad (6.56)$$

$$\left(\frac{c_\infty}{c}\right)^2 = 1 + \frac{R_m C'/C_p}{1 + \omega^2\tau'^2} \quad \cdots \cdots \quad (6.57)$$

$$\left(\frac{c}{c_0}\right)^2 = 1 + R_m(C'/C_p)\frac{\omega^2\tau'''^2}{(1 + \omega^2\tau'''^2)} \quad \cdot \quad \cdot \quad (6.58)$$

$$\frac{\alpha}{f^2} = \frac{1}{c_0}\frac{2\pi^2 R_m C'}{\sqrt{R_m C' C_p + C_p^2}}\frac{\tau'''}{(1 + \omega^2\tau'''^2)}\frac{1}{\sqrt{1 + \frac{R_m C'}{C_p}\frac{\omega^2\tau'''^2}{(1 + \omega^2\tau'''^2)}}}$$

$$(6.59)$$

$$\left(\frac{\alpha}{f^2}\right)\frac{c_0}{c} = \frac{2\pi^2}{c_0}\frac{R_m C'}{C_p + R_m C'}\frac{\tau'}{1 + \omega^2\tau'^2} \quad \cdot \quad \cdot \quad (6.60)$$

$$\left(\frac{\alpha}{f^2}\right)\frac{c}{c_0} = \frac{1}{c_0}\frac{2\pi^2 R_m C'}{\sqrt{R_m C' C_p + C_p^2}}\frac{\tau'''}{1 + \omega^2\tau'''^2} \quad \cdot \quad (6.61)$$

where:

$$\tau' = \frac{C_p - C'}{C_p}\tau_m \quad \cdot \quad \cdot \quad \cdot \quad \cdot \quad \cdot \quad (6.62)$$

$$\tau''' = \frac{C_p - C'}{\sqrt{C_p^2 + R_m C' C_p}}\tau_m \quad \cdot \quad \cdot \quad (6.63)$$

The terminology τ' and τ''' has been chosen to agree with Herzfeld and Litovitz's[6.3] usage.

Slightly different forms of the equations given above can be found in the literature. The differences arise from different methods used to solve equation (6.54). In particular, if we consider equation (6.58) for $(c/c_0)^2$ a different expression is obtained if we invert equation (6.54) and take the left-hand side which now has the form $(a - jb)^{-1}$ and write it as $\frac{1}{a}\left(1 + j\frac{b}{a}\right)$ from which the expression for $(c/c_0)^2$ originally derived by Kneser[6.4] is obtained as:

$$\left(\frac{c}{c_0}\right)^2 = 1 + R_m\left(\frac{C'}{C_p}\right)\frac{\omega^2\tau''^2}{1 + \omega^2\tau''^2} \quad \cdot \quad \cdot \quad (6.64)$$

where:

$$\tau'' = \left(\frac{C_v - C'}{C_v}\right)\tau_m$$

Typical examples of plots of equations (6.56) to (6.61) are given in Fig. 6.1.

6.7 Evaluation of Dispersion Curves

Provided that we have values for the velocity of ultrasonic waves in a fluid at differing frequencies we can use these in conjunction with the equations derived in the preceding section to obtain values for R_m, C' and the relaxation times from which we can obtain information about the molecular properties of the particular fluid.

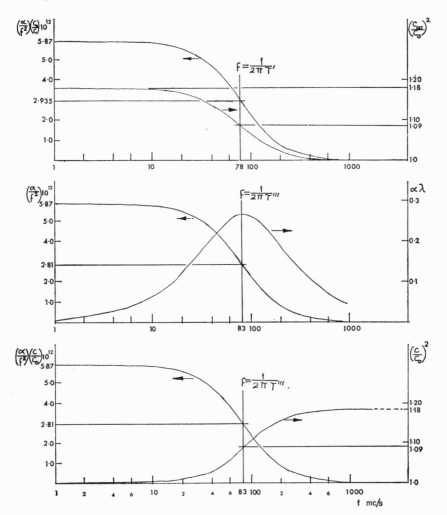

Fig. 6.1 Plots of equations 6.56 to 6.61 computed from the data for carbon disulphide obtained by Andrae, Heasell and Lamb (7.7).

If the complete dispersion curve is available so that we have values for c_∞ and c_0, we have, from either equation (6.57) or (6.58) that:

$$\left(\frac{c_\infty}{c_0}\right)^2 = 1 + R_m\frac{C'}{C_p} \qquad \qquad (6.65)$$

Provided that we know C_p we can find R_mC'. If R_m is known we can find C'. In order to find τ''' we consider the function $\dfrac{\omega^2\tau'''^2}{1 + \omega^2\tau'''^2}$. If this is plotted against $\log \omega\tau'''$ it has a point of inflection at $\omega\tau''' = 1$, where it has a value of 0·5. If instead of plotting $(c/c_0)^2$ against $\log \omega\tau'''$ we plot against $\log \omega$ we shall obtain the same curve shifted along the horizontal axis by $\log \tau'''$. If we can find the point of inflection we can find τ''', since the angular frequency at which this occurs is the reciprocal of τ'''.

If only just over half the complete dispersion curve is available the complete curve can be found by taking advantage of the symmetrical properties of $\dfrac{\omega^2\tau'''^2}{1+\omega^2\tau'''^2}$ when plotted as a function of $\log \omega$. Assuming that we have values for c/c_0 for frequencies below and slightly above the point of inflection we can complete the dispersion curve by taking the curve for $(c/c_0)^2 - 1$, plotted against $\log \omega$, rotating this curve about a horizontal line through the point of inflection by $180°$ and again rotating the resulting curve about a vertical line through the point of inflection by $180°$. We can then proceed as in the preceding paragraph.

If only the beginning of the dispersion curve is available and the velocity measurements are very accurate, we have that for $(\omega\tau''')^2 \ll 1$, equation (6.58) approximates to:

$$\frac{c - c_0}{c_0} \approx 0\cdot 5 R_m(C'/C_p)\omega^2\tau'''^2 \quad . \quad . \quad . \quad \textbf{(6.66)}$$

In this case we cannot obtain individual values for C', R_m and τ'''.

6.8 Evaluation of Absorption Curves

The absorption curves generally only provide values for R_m, C' and relaxation times when used in conjunction with dispersion data. If, for example, we combine equations (6.56), (6.58), (6.62) and (6.63) we obtain:

$$(c/c_0)^2 - 1 = 2\alpha c\tau' \quad . \quad . \quad . \quad . \quad \textbf{(6.67)}$$

Hence from values for dispersion and absorption at one frequency we can find τ'.

From equation (6.60) we have:

$$\alpha\frac{c_0}{c} = \tfrac{1}{2}\frac{1}{c_0\tau'} \cdot \frac{R_mC'}{C_p + R_mC'} \cdot \frac{\omega^2\tau'^2}{1+\omega^2\tau'^2} \quad . \quad . \quad \textbf{(6.68)}$$

Thus if we plot $\alpha c_0/c$ against $\log \omega$, the inflection point occurs at $\omega\tau' = 1$, hence we can find τ'. At the same frequency the value of $\dfrac{\alpha c_0}{c}$ is $\dfrac{1}{2\tau'c_0}\dfrac{R_mC'}{(C_p + R_mC')}$ hence we can find R_mC'.

If we combine equations (6.56) and (6.58) we have:

$$\alpha\lambda\frac{c_0^2}{c^2} = \frac{\pi R_mC'}{C_p} \cdot \frac{1}{1 + R_mC'/C_p} \cdot \frac{\omega\tau'}{1+\omega^2\tau'^2} \quad . \quad \textbf{(6.69)}$$

$\alpha\lambda c_0^2/c^2$ has a maximum when $\omega\tau' = 1$ where it equals $\dfrac{\pi R_mC'}{C_p(1 + R_mC'/C_p)}$ so that we can find τ' and R_mC'.

From equation (6.60) we have that the half value of $\dfrac{\alpha}{f^2}\dfrac{c_o}{c}$ occurs at $\omega\tau' = 1$ and is equal to $\dfrac{\pi^2 R_m C'}{C_p + R_m C'}$. Hence we can find $R_m C'$ and τ'.

While the preceding techniques for evaluating the absorption data also require dispersion data there are two cases where a knowledge of dispersion is not required.

If the internal specific heat C' is small compared with C_p then equation (6.60) becomes:

$$\frac{\alpha}{\omega} \approx \frac{1}{2c_o}\frac{R_m C'}{C_p + R_m C'}\frac{\omega\tau'}{1 + \omega^2\tau'^2} \quad \cdot \quad \cdot \quad \cdot \quad \textbf{(6.70)}$$

α/ω has a maximum when $\omega\tau' = 1$, hence we can find τ' and $R_m C'$ provided that we know c_o.

If, in addition to C' being much smaller than C_p, $\omega\tau' \ll 1$, equation (6.70) becomes:

$$\alpha \approx \frac{1}{2c_o}\frac{R_m C'}{C_p + R_m C'}\omega\tau' \quad \cdot \quad \cdot \quad \cdot \quad \cdot \quad \textbf{(6.71)}$$

Thus we can find $R_m C'$ or τ' provided that the other is known.

The expressions given above for absorption and dispersion are valid for both structural and thermal relaxation processes. These general expressions enable us to find 'relaxation times' but not the relaxation time of the process or C' unless R_m is known. If, however, structural processes are absent and only thermal processes exist, as is the case for gases and many liquids, $\Delta E_m = \Delta H_m$ so that now:

$$R_m = \frac{C_p - C_v}{C_v - C'} \quad \cdot \quad \cdot \quad \cdot \quad \cdot \quad \textbf{(6.72)}$$

Knowing C_p and C_v and $R_m C'$ now enables us to find C' and so if we also know τ' or τ''' we can find τ_m by means of equations (6.62) or (6.63).

References

6.1 Andrae, J. H. and Lamb, J., *Proc. Phys. Soc.*, B, **69**, 814 (1956).
6.2 Callen, H. B., *Thermodynamics*, p. 118. Wiley (1960).
6.3 Herzfeld, K. F. and Litovitz, T. A. *Absorption and Dispersion of Ultrasonic Waves*, Chap. 2. Academic Press (1959).
6.4 Kneser, H. O., *Ann. Phys.* (Leipzig), **11**, 761 (1931).

7
Absorption and Dispersion in Gases and Liquids

7.1 Introduction

In the preceding chapter we developed a formal theory for the absorption and dispersion of ultrasonic waves by assuming the existence of some relaxation process in addition to the classical processes of viscosity and thermal conductivity. We did not, however, consider the actual molecular mechanisms involved. In this chapter we will be considering, albeit briefly, the various molecular mechanisms so far identified. In comparing experimental with calculated values, the absorption coefficients are usually expressed in terms of α/f^2 since at frequencies some way below that at which $\omega\tau$ tends to unity equations (6.11), (6.20) and (6.59) show that α/f^2 is constant.

7.2 Gases

The experimental values for α/f^2 for gases do not differ markedly from those calculated from viscosity and thermal conductivity data except in a few cases and near relaxation frequencies. For example, the experimental value of α/f^2 for nitrogen[7.1] at 20° C and one atmosphere pressure is $1\cdot58$ 10^{-11} sec^2 m^{-1} whereas the calculated components are 0·96 10^{-11} sec^2 m^{-1} and 0·39 10^{-11} sec^2 m^{-1} for viscosity and thermal conductivity respectively. Thus the ratio of experimental to classical values is about 1·24. The majority of the excess absorption of 0·23 10^{-13} is due to the slow energy exchange between the translational and rotational degrees of freedom of the nitrogen molecules with a slight contribution from the vibrational energy.

The relaxation process responsible for the excess absorption of a gas over the classical value is the slow rate of energy interchange between the external or translational degrees of freedom and the internal, rotational and vibrational, degrees of freedom. At equilibrium at a given temperature the energy of a gas molecule is distributed in definite proportions between the internal and external degrees of freedom. The transfer of energy between one degree of freedom and another only occurs when the gas molecule collides with another molecule. If now we

128

suppose that the gas temperature is raised suddenly, the molecule's translational energy will increase immediately. The energy distribution, however, between the degrees of freedom will no longer be in equilibrium and the molecule will have to suffer a number of collisions before equilibrium is re-established. This picture indicates how the specific heat of a gas can have differing instantaneous and equilibrium values. If we, for example, inject a given quantity of heat into a gas all the heat energy appears initially as translational energy corresponding to a particular temperature rise. As some of the heat energy starts to appear as rotational and vibrational energy the translational energy and therefore the temperature rise must decrease. Thus the effective specific heat of the gas increases with time and we can write:

$$C_{\text{effective}} = C_{\text{translational}} + C_{\text{internal}} \qquad . \quad . \quad (7.1)$$

where C_{internal} is a function of time (cf. equations 6.35 and 6.52).

To a first approximation the specific heat of a gas at constant volume is composed of independent contributions from each degree of freedom. If we consider the molecular specific heat at constant volume, $\bar{C}_v$, each of the translational and rotational degrees of freedom contributes[7.2] $R/2$ while each vibrational degree of freedom contributes $R(T_v/2T)^2/\sinh^2(T_v/2T)$ where R denotes the gas constant per mole, T the absolute temperature of the gas and T_v a characteristic vibrational temperature. Thus we have:

For a monatomic gas:

$$\bar{C}_v = 1{\cdot}5R \qquad . \quad . \quad . \quad . \quad . \quad (7.2)$$

For a linear molecule with n atoms:

$$\bar{C}_v = \underset{\text{translation}}{1{\cdot}5R} + \underset{\text{rotation}}{2(R/2)} + \underset{\text{vibration}}{(3n-5)}\frac{R(T_v/2T)^2}{\sinh^2(T_v/2T)} \quad . \quad (7.3)$$

For a non-linear molecule of n atoms:

$$\bar{C}_v = \underset{\text{translation}}{1{\cdot}5R} + \underset{\text{rotation}}{R} + \underset{\text{vibration}}{(3n-6)}\frac{R(T_v/2T)^2}{\sinh^2(T_v/2T)} \quad . \quad (7.4)$$

We can use the expressions above for $\bar{C}_v$ in the equations developed in chapter 6 to find expressions for the absorption and dispersion for a gas. If we consider a mole of gas we have:

$$PV = RT \qquad . \quad . \quad . \quad . \quad . \quad (7.5)$$

Combining equations (6.12), (6.15) and (7.5) gives the compressibility as:

$$K = 1/\gamma P$$

Also, since the density ρ is given by:

$$\rho = M/V$$

where M denotes the molecular weight of the gas, we find c_0, from equation (1.24), to be given by:

$$c_0{}^2 = \gamma RT/M \ . \quad . \quad . \quad . \quad . \quad (7.6)$$

The thermal conductivity K_h is given fairly accurately by [7.3]:

$$K_h = \tfrac{1}{4}(9\gamma - 5)\eta\frac{\bar{C}_v}{M} \quad . \quad . \quad . \quad . \quad (7.7)$$

Finally, for an ideal gas we have:

$$\bar{C}_p - \bar{C}_v = R \quad . \quad . \quad . \quad . \quad (7.8)$$

We can now substitute the preceding equation into equations (6.10), (6.11), (6.20) and (6.21) to find the total classical absorption and dispersion, viz.:

$$\alpha_{\text{class.}} = \tfrac{2}{3}\frac{\omega^2}{c^2\rho}\eta\left[1 + \tfrac{3}{4}\frac{R}{C_v}\left(1 + \frac{5}{4}\frac{R}{C_p}\right)\right] \quad . \quad . \quad (7.9)$$

$$\left(\frac{c_0}{c}\right)^2_{\text{class.}} = 1 - \tfrac{4}{9}\frac{\omega^2}{c^2\rho^2}\eta^2\left\{\left[1 - \tfrac{3}{16}\left(\frac{\gamma - 1}{\gamma}\right)(9\gamma - 5)\right]^2 - \right.$$
$$\left. - \tfrac{9}{64}\left(\frac{\gamma - 1}{\gamma^2}\right)(9\gamma - 5)^2\right\} \quad (7.10)$$

7.3 Non-classical Absorption in Gases

If we take the relaxing component of the molecular specific heat, $\bar{C}'$, to be that for vibrations and denote the number of rotational degrees of freedom by r, then:

$$\bar{C}_v = (3 + r)R_2 2 + \bar{C}'$$

whence:

$$\alpha\lambda = \frac{4\pi\bar{C}'}{R\left\{\left(5 + r + \dfrac{2\bar{C}'}{R}\right)\left(3 + r + \dfrac{2\bar{C}'}{R}\right)(5 + r)(3 + r)\right\}^{\frac{1}{2}}}$$
$$\left(\frac{\omega\tau''}{1 + \omega^2\tau''^2}\right) \quad (7.11)$$

$$\left(\frac{c}{c_0}\right)^2 = 1 + \frac{4\bar{C}'}{(3 + r)[(5 + r)R + 2\bar{C}']}\left(\frac{\omega^2\tau''^2}{1 + \omega^2\tau''^2}\right) \quad (7.12)$$

$$\tau'' = \frac{(5 + r)^{\frac{1}{2}}}{2\left[\left(5 + r + \dfrac{2\bar{C}'}{R}\right)\left(3 + r + \dfrac{2\bar{C}'}{R}\right)(3 + r)\right]}\tau_m \quad (7.13)$$

7.4 Relation between Collision Frequency and Relaxation Time

We stated in section 7.2 that the transfer of energy between the degrees of freedom only occurs during inelastic collisions between molecules. If we assume that only binary collisions are necessary, that is collisions involving only two molecules, we would expect the relaxation time to be inversely proportional to the number of collisions occurring per second made by any one molecule, or alternatively expressed, to be proportional to, and of the same order as, the average time interval, τ_c, between collisions. However, not all collisions promote an energy interchange so that the relaxation time, τ_s, for the particular internal degree of freedom will be greater than τ_c. Thus we may write:

$$\tau_s = Z_s \tau_c \qquad \qquad \text{(7.14)}$$

where Z_s in fact denotes the number of collisions necessary to reduce the excess or lack of internal energy relative to its equilibrium value by a factor of 2·718 (e). Z_s is termed the effective collision number and may be found from the ratio of the total to the classical absorption at frequencies well below those for which $\omega\tau_s = 1$ provided that the relaxation times for rotation and vibration are widely separated. The viscosity of a gas at a pressure p is given by[7.4]:

$$\eta = 1\cdot271 p \tau_c \qquad \qquad \text{(7.15)}$$

and if this expression is substituted into equation (7.9) which is then combined with equation (6.71) suitably modified to describe gases we find:

$$\frac{\alpha}{\alpha_{\text{class.}}} = 1 + 0\cdot0671\left(Z_r + \frac{\bar{C}'}{R}Z_v\right) \quad \text{for linear molecules}$$

$$\frac{\alpha}{\alpha_{\text{class.}}} = 1 + 0\cdot074\left(Z_r + \tfrac{2}{3}\frac{\bar{C}'}{R}Z_v\right) \quad \text{for non-linear molecules}$$

where Z_r and Z_v are the effective collision numbers for rotation and vibration respectively and $\bar{C}'$ the vibrational molecular specific heat. Typical values for Z_r and Z_v at room temperature are 300 and 10^6 respectively.

The assumption that the redistribution of energy is effected through collisions leads to one important consequence in that since τ_s is proportional to τ_c and as, at a given temperature τ_c is inversely proportional to density ρ, we would expect $\tau_s\rho$ to be constant. Provided that the gas pressure is not excessive the gas density is directly proportional to pressure p so that we would expect $\tau_s p$ to be constant. Thus denoting the relaxation time at a standard pressure p_0 by τ_0 we have:

$$\tau_0 p_0 = \tau_s p$$

whence:

$$\omega\tau_s = \frac{\omega}{p}(p_0\tau_0) \quad . \quad . \quad . \quad . \quad . \quad \text{(7.18)}$$

Equation (7.18) is important experimentally since the expressions for absorption and dispersion derived in chapter 6 are functions of $\omega\tau$ which for gases may be replaced by ω/p. Thus we can obtain absorption or dispersion curves for gases by varying the static pressure as well as the ultrasonic frequency and plotting the results against ω/p as illustrated in Fig. 7.1. For maximum accuracy corrections for the non-ideality of real gases should be made.

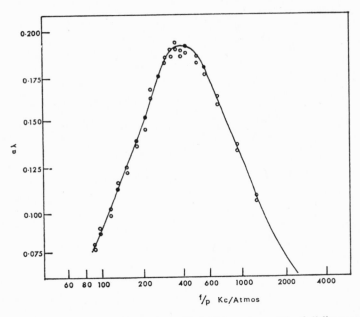

Fig. 7.1 Experimental results for carbon disulphide vapour. The full line represents the theoretical curve assuming a single relaxation. (After Angona (8.1).)

7.5 Gas Mixtures

Early measurements of the absorption of supposedly pure gases gave widely varying results. This variation was traced to the presence of differing concentrations of impurities in the gases since even a small amount of impurity can lower the absorption to a marked extent. It appears that the impurity molecules can be more effective in promoting energy redistribution in the gas molecules than molecules of the same gas. Since liquid mixtures however have been more extensively studied than gaseous ones we will leave a discussion of the absorption of mixtures until we deal with liquids.

7.6 Liquids

While one mechanism leads to absorption in gases four have so far been identified for liquids. Before we consider these it is convenient to divide liquids into groups according to the ratio of measured to classical absorptions and the temperature coefficient of absorption. This classification, originally given by Pinkerton[7.5], is given in Table 7.1.

TABLE 7.1

Group	α/α class	$d\alpha/dT$	Type of Liquid	Examples
1	3–1500	positive	unassociated polyatomic	CS_2, CCl_4, C_6H_6
2	1·5–3	negative	associated polyatomic	H_2O, CH_3OH
3	5–5000	positive or negative	organic acids	CH_3COOH
4	1	positive	esters monatomic	$CH_3COOC_2H_5$ Hg, liquid A,
5	1	negative	diatomic associated polyatomic	liquid O_2, H_2 highly viscous liquids

Liquids belonging to the first three groups all show absorption coefficients greater than the classical values so that we would expect to find relaxation processes occurring. Since the temperature coefficients of the absorption coefficients vary between groups the existence of differing mechanisms is indicated although this evidence is not conclusive since a temperature coefficient is usually the resultant of several competing effects. Liquids belonging to group 4 show no relaxation while those in group 5 could often be reclassified into group 2.

The absorption shown by liquids in group 1 is produced by the same mechanism as that for gases and they are usually referred to as Kneser liquids after H. O. Kneser[7.6] who first gave the theory of thermal relaxation. As an example of a Kneser liquid we consider carbon disulphide which has been thoroughly investigated by Andrae, Heasell and Lamb[7.7] over a frequency range of from 6 mc/s to 190 mc/s. These workers assumed that the absorption is due to the relaxation of the entire vibrational specific heat occurring at constant volume so that $\Delta E_m = \Delta H_m$. Denoting τ''' by $1/f_c$, equations (6.72), (6.58) and (6.60) give:

$$\left(\frac{c}{c_o}\right)^2 = 1 + \left(\frac{\bar{C}_p - \bar{C}_v}{\bar{C}_v - \bar{C}'}\right)\left(\frac{\bar{C}'}{\bar{C}_p}\right)\frac{1}{1 + (f_c/f)^2} \quad . \quad (7.19)$$

$$\frac{\alpha}{f^2}c = \frac{A}{1 + (f/f_c)^2} + B \quad . \quad . \quad . \quad . \quad (7.20)$$

where

$$A = \frac{2\pi^2(\bar{C}_p - \bar{C}_v)\bar{C}'}{[(\bar{C}_v - \bar{C}')(\bar{C}_p - \bar{C}')\bar{C}_p\bar{C}_v]^{\frac{1}{2}}}$$

and B denotes the contribution to $\frac{\alpha}{f^2}c$ from other absorption processes such as those with relaxation frequencies differing greatly from f_c and the classical mechanisms. The contribution from B may be neglected since it is of the order of 1% of the minimum measured value of $\frac{\alpha}{f^2}c$.

Andrae, Heasell and Lamb found values for A and f_c by trying various values until the calculated and experimental values of α/f^2 agreed and their results are shown in Fig. 7.2. Taking $\bar{C}_p = 18\cdot17$, $\bar{C}_v = 11\cdot71$ and

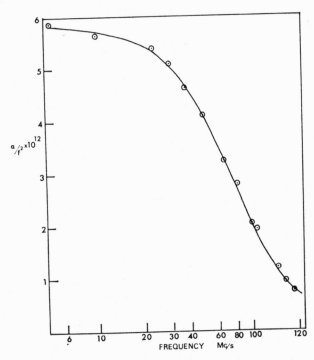

Fig. 7.2 Andrae, Heasell and Lamb's data for carbon disulphide compared with their best fitting computed values represented by the solid curve (7.7).

$\bar{C}' = 3\cdot933$ cal mole^{-1} deg^{-1} they found $A = 6.7\ 10^{-9}$ secs, $f_c = 78$ mc/s and $\tau_m = 2.825\ 10^{-9}$ secs. The value of τ_m for liquid CS_2 should be compared with that for gaseous CS_2 which at about the same temperature is $7\ 10^{-7}$ secs. These figures illustrate a general principle that the relaxation times for liquids are two or three orders of magnitude smaller than for gases at atmospheric pressure as we would expect since the liquid density is considerably greater than that of the gas.

7.7 Rotational Isomerism

Many chemical compounds can exist in two or more isomeric forms, that is molecules with the same chemical formula can possess differing geometrical structures arising from rotations of groups about their valence bonds. For example, acraldehyde can exist in the two forms shown in Fig. 7.3. At any given temperature an equilibrium exists

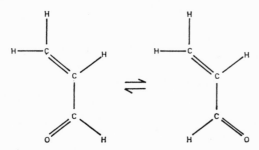

Fig. 7.3 Isomeric forms of acraldehyde.

between the two isomers. This equilibrium is perturbed by the passage of an ultrasonic wave and if the isomers possess differing energies the transfer between them will proceed at a definite rate, that is we have a relaxation process which will give rise to absorption of the ultrasonic energy. The variation in ultrasonic properties with temperature allows us to measure the activation energies for the transfer from one isomer to another. The free energy diagram for the reaction between two isomeric forms is shown in Fig. 7.4 in which the terms k_{12} and k_{21} denote the

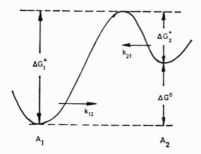

Fig. 7.4 Free energy diagram for the reaction between two isomers.

reaction rates for the transfer from isomer A_1 to A_2 and from A_2 to A_1 respectively. The absorption can be represented by equation (7.21) which has been derived from equation (7.20) under the assumption that dispersion is negligible as is usually the case for isomeric relaxation.

$$\frac{\alpha}{f^2} = \frac{A'}{1 + (f/f_c)^2} + B' . \quad . \quad . \quad . \quad . \quad (7.21)$$

By assuming that:

$$f_c = (k_{12} + k_{12})/2\pi$$
$$k_{12} \ll k_{21}$$
$$\Delta G° > 3RT$$

where $\Delta G°$ is the Gibbs free energy difference between the two isomers, Lamb[7.8] obtains the following expressions:

$$\frac{2(\alpha\lambda)_m}{\pi} = \frac{c^2}{J}\left(\frac{\theta}{C_p}\right)\frac{(\Delta H°)^2}{RT}e^{-\Delta H°/RT}\, e^{\Delta S°/R} \quad . \quad . \quad (7.22)$$

$$f_c = \frac{F_2}{2\pi}\left[\frac{kT}{h}\right]e^{-\Delta H_2^+/RT} \quad . \quad . \quad . \quad . \quad (7.23)$$

where $(\alpha\lambda)_m$ denotes the maximum value of the absorption per wavelength, θ the coefficient of expansion of the liquid, $J = 4\cdot186\ 10^{-7}$ erg cal^{-1}, $\Delta S°$ the entropy difference between the isomers, k Boltzmann's constant, h Plank's constant and F_2 a numerical factor representing the entropy of activation for the transition $A_2 \rightarrow A_1$. If we plot $\log_e T(\alpha\lambda)_m/c^2$ against T^{-1} equation (7.22) shows that we should obtain a straight line of slope equal to $\Delta H°/R$. Similarly if we plot $\log_e(f_c/T)$ against T^{-1} equation (7.23) shows that we should obtain a straight line of slope equal to $\Delta H_2^+/R$. Thus from ultrasonic measurements we can obtain activation energies and energy differences for the two isomeric forms.

7.8 Monomer–Dimer Reactions

In some polar liquids, such as acetic and propionic acids, an equilibrium between monomers and dimers exists and ultrasonic absorption arises from the finite time required for the equilibrium to be re-established

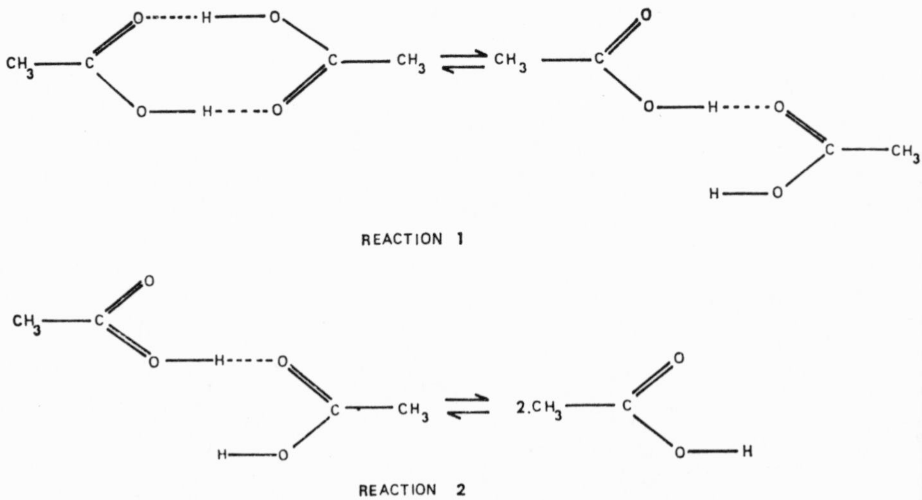

REACTION 1

REACTION 2

Fig. 7.5 Configurations taken up by acetic acid. Reaction 1 is that responsible for ultrasonic absorption according to Tabuchi (7.11).

after it has been disturbed. Acetic acid has been extensively studied by Lamb and Pinkerton[7.9], Freedman[7.10] and Tabuchi[7.11]. According to Tabuchi the reaction takes place in two stages illustrated in Fig. 7.5 where the first reaction is the one responsible for ultrasonic absorption.

7.9 Structural Relaxation

All the relaxation mechanisms so far discussed arise essentially from the perturbation of an equilibrium caused by the small temperature changes accompanying the passage of an ultrasonic wave through a liquid. However, as well as temperature changes, pressure changes also occur. If a liquid's molecules form loose structures and differing structures have differing energies and the differing structures are in equilibrium at a given pressure, a pressure change may shift the equilibrium. The new equilibrium will take a certain time to establish itself and hence we have a relaxation effect which can lead to ultrasonic absorption. Hall[7.12] has applied this idea to account for the absorption of water. He assumes that water is composed of two groupings of molecules, one consisting of close-packed and one of open-packed molecules. Unfortunately the calculated relaxation time is $3\cdot5 \ 10^{-12}$ secs which makes a direct check on the theory not yet feasible.

Structural mechanisms are also thought to be responsible for the excess absorption of other associated and highly viscous liquids but little direct evidence is yet available.

7.10 Mixtures

We mentioned in section 7.5 that the absorption of a gas containing a small amount of impurity can be considerably lower than for the pure gas. The same effect is found in liquid mixtures. If a Kneser liquid, A, shows a high absorption the intermolecular collisions must be relatively inefficient in causing the transfer of energy between the various degrees of freedom while for a liquid, B, with low absorption the collisions must be relatively efficient. If we now add a small amount of B to A the energy transfer efficiency for type A molecules when they collide with type B will be increased so that the mixture will display a lower absorption than for pure A. The absorption reduction will in fact be greater than would be expected on a simple proportionality basis—this is illustrated in Fig. 7.6. According to Sette[7.13, 7.14] the absorption of a mixture containing a mole fraction x of B is given by:

$$\frac{\alpha}{f^2} = \left(\frac{\alpha}{f^2}\right)_B \left\{ 1 - \left[1 - \left(\frac{c\bar{C}_v\bar{C}_p}{\bar{C}_p - \bar{C}_v}\right)_B \left(\frac{\bar{C}_p - \bar{C}_v}{c\bar{C}_v\bar{C}_p}\right)_A \right] \right\} \cdot$$
$$\left\{ x \frac{\tau_{AA}}{\tau_{BB}} \frac{\bar{C}'_A}{\bar{C}'_B} \left[x + (1-x)\frac{\tau_{AA}}{\tau_{AB}} \right]^{-1} + (1-x)\left[1 - x + x\frac{\tau_{BB}}{\tau_{BA}} \right]^{-1} \right\}$$

$$\text{(7.24)}$$

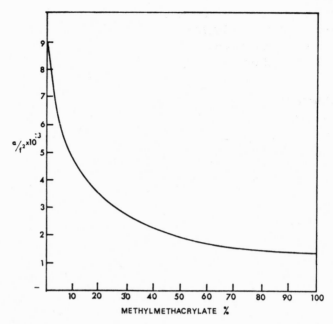

Fig. 7.6 Absorption of a mixture of methylmethacrylate and benzene.

where τ_{AA} denotes the relaxation time of pure A, τ_{BB} that for pure B, τ_{AB} that for the deactivation of A by B and τ_{BA} that for the deactivation of B by A.

References

7.1   Parker, J. G., Adams, C. E. and Stavseth, R. M., *J. Acoust. Soc. Amer.*, **25**, 263 (1953).
7.2   Callen, H. B., *Thermodynamics*, p. 331. Wiley (1960).
7.3   Euken, A., *Phys. Zeits.*, **14**, 324 (1913).
7.4   Glasstone, S., *Elements of Physical Chemistry*, p. 97. Macmillan (1950).
7.5   Pinkerton, J. M. M., *Proc. Phys. Soc.*, B, **62**, 129 (1949).
7.6   Kneser, H. O., *Ann. Phys.* (5), **16**, 337 (1933).
7.7   Andrae, J. H., Heasell, E. L. and Lamb, J., *Proc. Phys. Soc.*, B, **69**, 625 (1956).
7.8   Lamb, J., *Z. f. Elektrochem.*, **64**, 135 (1960).
7.9   Lamb, J. and Pinkerton, J. M. M., *Proc. Roy. Soc.*, A, **199**, 114 (1949).
7.10 Freedman, A., *J. Chem. Phys.*, **21**, 1784 (1953).
7.11 Tabuchi, D., *Proc. 3rd Int. Cong. Acoust.*, p. 500 (1959).
7.12 Hall, L., *Phys. Rev.*, **73**, 772 (1948).
7.13 Sette, D., *J. Chem. Phys.*, **18**, 1592 (1950).
7.14 Sette, D., *J. Acoust. Soc. Amer.*, **23**, 359 (1951).

8

Measurement Techniques for Longitudinal Waves

8.1 Introduction

In this chapter we shall be concerned with the various techniques employed to measure the absorption and velocity of longitudinal waves. We shall first consider those methods applicable to gases only and secondly those methods applicable to both gases and liquids. Some of these methods are also applicable to solids.

8.2 Travelling Wave Tube

In this method due to Angona[8.1] plane waves are generated by means of a small ribbon loudspeaker situated within a long glass tube whose ends are plugged with glass wool to eliminate standing waves. A condenser microphone is attached to the side of the tube and the amplitude of the signal from the microphone is monitored as the loudspeaker is moved along the inside of the tube by means of an external magnet. The decrease in microphone output as the loudspeaker is moved away gives directly the energy loss within the tube as a function of distance. This energy loss arises from both the absorption of the gas and the wall losses arising from viscous and thermal effects at the tube walls. These latter losses can be calculated but they are better determined by calibrating the system with a known gas. The accuracy of this method lies between 2·5% and 10% depending upon the gas absorption coefficient. It is possible to make velocity measurements by noting the distance travelled by the loudspeaker between positions along the tube for which the microphone output signal has the same phase relative to the loudspeaker excitation signal.

8.3 Reverberation

A column of gas enclosed in a tube with plane parallel ends behaves as an acoustic transmission line terminated in high impedances so that if the gas column is an integral number of half-wavelengths long it will resonate. If the acoustic excitation at a resonant frequency is applied and then removed the acoustic amplitude at any point within the tube will decay exponentially at a rate governed by the energy losses of the system. Since these arise from the gas absorption and the wall losses the former can be found from a measurement of the total loss derived from the rate of decay if the wall losses are known. Edmonds and Lamb[8.2] have used this technique to measure absorption coefficients of gases over the range of 3 to 20 kc/atmos. In their equipment, illustrated in Fig. 8.1,

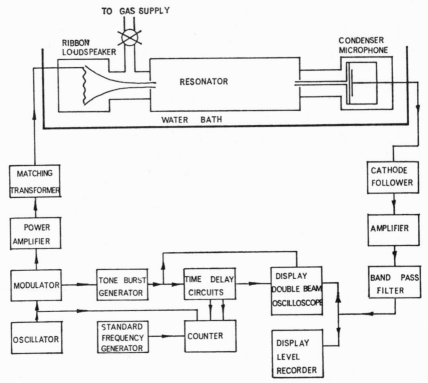

Fig. 8.1 Apparatus used by Edmonds and Lamb (8.2) for measuring the absorption of gases.

the gas was enclosed in a precision glass tube, 60 cm long by 9·8 cm diameter, immersed in a thermostatted water bath. The gas was excited acoustically by a ribbon loudspeaker which was driven for a short while at one of the column's resonant frequencies. When the loudspeaker was switched off the decaying sound amplitude within the gas was monitored by a condenser microphone and the resulting waveform displayed on an oscilloscope. The gain of the amplifier between the microphone and the oscilloscope, together with the time base speed, were varied until the exponentially decaying waveform envelope matched up with one drawn on the face of the cathode ray tube. Two time markers were then moved along the waveform until they lined up with two points so arranged that the ratio of the microphone signals at these points was $e : 1$. The time interval between these markers was then measured by an accurate crystal controlled time interval meter. If the measured time interval is denoted by T_m, Edmonds and Lamb show that:

$$T_m = 1/\alpha_m c \quad . \quad . \quad . \quad . \quad . \quad \textbf{(8.1)}$$

where α_m and c denote the total absorption coefficient and acoustic velocity respectively. The absorption coefficient due to thermal and viscous losses at the walls can be calculated with reasonable accuracy provided that the particular mode, longitudinal or radial, being excited is known. Denoting these losses by an absorption coefficient α_c and assuming the losses to be simply additive we have:

$$\alpha_{\text{gas}} = \alpha_m - \alpha_c \quad . \quad . \quad . \quad . \quad \textbf{(8.2)}$$

where α_{gas} denotes the required absorption coefficient for the gas. In order to use equation (8.1) to find α_m the acoustic velocity c must be known. If this is not available from other measurements it can be found from the resonant frequency of a longitudinal mode. Edmonds and Lamb claim that their system will give results accurate to about 5%.

8.4 Harlow's Method

Harlow[8.3] contained his gas in a cylindrical tube terminated at one end by a plane rigid reflector. The other end of the tube contained an electro-magnetically driven piston. As the driving frequency of the piston passes through a resonant frequency of the gas system the apparent electrical impedance of the piston drive is monitored from which the acoustic velocity and absorption of the gas in the tube can be derived. The electrical impedance of the piston, Z, can be represented by:

$$Z = Z_p + (R + jX) \quad . \quad . \quad . \quad . \quad \textbf{(8.3)}$$

where Z_p denotes the impedance of the piston alone and the other terms denote the acoustic impedance at the face of the piston multiplied by the piston's transformation factor (cf. the factor $h^2 C_0^2$ for a piezoelectric transducer). According to equation (1.51) the acoustic impedance at the face of the piston looking into a column of gas of length l terminated in an infinite impedance is given by:

$$Z_0 \frac{1 + j \tanh\alpha l \,.\, \tan \beta l}{\tanh\alpha l + j \tan \beta l}$$

The phase angle of this impedance, θ, is given by:

$$\tan \theta = \frac{X}{R} = \tfrac{1}{2} \left[\frac{\tanh^2\alpha l - 1}{\tanh\alpha l} \right] \sin 2\beta l \quad . \quad . \quad \textbf{(8.4)}$$

whence, provided that $\alpha l \ll 1$,

$$\left.\frac{d(\tan \theta)}{df}\right|_{\tan \theta = 0} = S = -\frac{2\pi}{c\alpha} \qquad \ldots \quad \text{(8.5)}$$

that is:

$$\alpha = -2\pi/Sc \quad \ldots \quad \ldots \quad \text{(8.6)}$$

The velocity of sound in the gas in the tube can be found from the fact that when $\tan \theta = 0$, $\sin 2\beta l = 0$, hence:

$$c = 2fl/n \quad \ldots \quad \ldots \quad \ldots \quad \text{(8.7)}$$

where n is an integer. Thus from equations (8.6) and (8.7) we can find the absorption and velocity for the gas in the tube.

In order to find $R + jX$ from the measured piston impedance Z we require to know Z_p. This can be found by measuring Z at a frequency midway between two frequencies at which $\tan \theta = 0$ since then $\tan \beta l$ becomes infinitely large and the acoustic impedance at the face of the piston, and hence $R + jX$, becomes negligibly small.

With the diameter of tube and frequencies used by Harlow the ratio of diameter to wavelength is very small. Under this condition the measured sound velocity differs from the free space value and the absorption is due mainly to viscous and thermal losses at the boundary between the gas and the tube wall. If we denote the free space velocity by c_0 and the tube radius by r the following conditions[8.4] apply:

$$c_0 = c\left[1 + \frac{c_0}{c}\frac{1}{r}\sqrt{\frac{\nu}{2\omega}}\right] \qquad \ldots \quad \text{(8.8)}$$

$$\alpha = \frac{1}{rc_0}\sqrt{\frac{\omega\nu}{2}} \qquad \ldots \quad \ldots \quad \text{(8.9)}$$

where ν is a constant for the gas depending upon its viscosity and thermal conductivity. Combining equations (8.8) and (8.9) gives:

$$c_0 = c\left[1 + \frac{c_0}{c}\frac{\alpha\lambda_0}{2\pi}\right] \quad \ldots \quad \ldots \quad \text{(8.10)}$$

where λ_0 denotes the free space wavelength. Thus equation (8.10) coupled with a knowledge of c and α enables us to find the free space acoustic velocity for the gas. Harlow claims to be able to measure c to an accuracy of one part in 30,000 at frequencies around 1 kc/s.

If the gas shows appreciable absorption due to relaxation processes this method will require modification since then α will be the sum of two components; one corresponding to equation (8.9) and one to relaxation. Assuming that these are additive the absorption due to boundary layer losses can be derived by plotting the absorption measured at the same frequency but in tubes of differing radii against $1/r$ when the intercept at $1/r = 0$ will give the absorption due to relaxation. If this is subtracted from the total absorption the boundary layer absorption is found and can then be used with equation (8.10) to give the free space velocity.

8.5 Reverberation (Liquids)

The reverberation technique as applied to liquids is, in essence, identical to that used for gases. Karpovitch[8.5] carried out extensive investigations into the reverberation technique and his system is illustrated in Fig. 8.2.

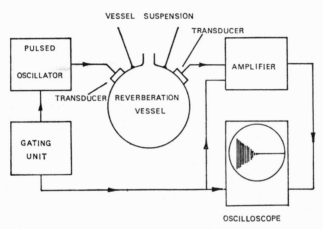

Fig. 8.2 Apparatus used by Karpovitch (8.5) for measuring the absorption of liquids by means of reverberation.

A half litre spherical vessel was completely filled with the liquid under investigation and suspended by fine wires. Two transducers were attached to the vessel, one to excite oscillations and one to record their decay when the excitation was removed. A pulsed sinusoidal voltage was applied to the exciting transducer and at the end of the pulse the amplifier coupling the second transducer to the oscilloscope was switched on to pass the decaying signal from the transducer into the oscilloscope for display. The rate of decay of the oscillations of the liquid in the sphere was measured in a similar manner to that used by Edmonds and described in section 8.3 and the liquid's attenuation was derived from the rate of decay also as before. Karpovitch assumed that only radial modes were excited since he chose excitation frequencies which produced the longest and most exponential decays. He assumed that the total losses of the system would be minimal for radial modes since with these no motion of the fluid occurred tangentially to the inner surface of the container. The measured losses in such a system are the sum of the absorption losses in the liquid plus those arising from the motion of the vessel and its suspension. In order to correct for these Karpovitch calibrated his system with liquids of known absorption covering a range of absorptions and frequencies since he found that the required correction was a function of both frequency and liquid absorption. With the correction applied, Karpovitch estimated his accuracy to be about 5%. More recently Andrae and Edmonds[8.6] have cast some doubt upon Karpovitch's assumption that only radial modes were excited since they found that by using extremely accurate spherical quartz resonators the radial mode could only be excited if the sphere were supported on an air bearing and inside a continuously evacuated chamber.

, Reverberation techniques allow measurements at frequencies down to a few kilocycles/second. No detailed measurements have yet been made at much lower frequencies but Andrae and Edmonds[8.6] carried out some preliminary work using a quartz tuning-fork, containing the liquid in the bridge connecting the fork arms, oscillating at about 120 c/s. This method should enable measurements of the adiabatic compressibility of the liquid to be made by noting the difference in the fork's resonant frequency with and without the liquid. A measurement of the rate of decay of the fork's vibrations should give the liquid's absorption. As with Karpovitch's system the fork requires calibration with liquids of known properties.

8.6 Streaming (Liquids)

In section 2.8 we showed that if a liquid absorbs ultrasonic radiation a radiation pressure is set up which can cause the liquid to flow and according to equation (2.44) the volume flow can be used to obtain the liquid's absorption coefficient. We consider a system illustrated in Fig. 8.3 in which a transducer radiates into a column of liquid from which it

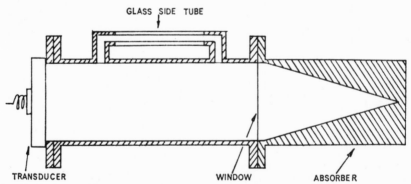

Fig. 8.3 Apparatus for the measurement of the absorption of liquids using streaming.

passes via a window into a suitably highly absorbing medium so that no reflected waves occur. If we attach a side tube to the main tube the radiation pressure difference Δp between the two ends of this side tube is given by:

$$\Delta p = E[e^{-2\alpha x_1} - e^{-2\alpha x_2}] \qquad . \quad . \quad . \quad \textbf{(8.11)}$$

where E denotes the radiation energy density at the face of the transducer. If the side tube has a radius r and length l and the liquid a viscosity η, the pressure difference Δp will cause the liquid in the side tube to flow with a velocity v given by Poiseuille's law as:

$$v = \frac{\Delta p r^2}{4\eta l} \qquad \qquad \textbf{(8.12)}$$

If we now combine equations (8.11) and (8.12) and introduce a calibration constant k to account for the end effects associated with the connections of the side tube to the main tube we find:

$$\frac{k\eta v}{E} = e^{-2\alpha x_1} - e^{-2\alpha x_2} \qquad \qquad \textbf{(8.13)}$$

Thus provided we can find k, v and E we can obtain a value for the absorption coefficient. The correction factor k can be found by calibrating the system with liquids of known absorptions. The velocity v can be measured by timing the passage over a known distance of small aluminium particles suspended in the liquid in the side tube and observed through a travelling microscope.

Two methods have been developed for measuring the energy density E at the face of the transducer. Piercy and Lamb[8.7] placed a reflector at 45° to the direction of travel of the sound wave so that the radiation was turned through 90° before passing into the absorber. The reflector took the form of a paddle consisting of two thin metal plates separated at their edges by an elliptical ring so that the space between the plates contained air only. The paddle was suspended by a fine torsion wire. By noting the deflection of the paddle when it was reflecting the acoustic radiation the radiation pressure at the front surface of the paddle could be calculated and hence one could work back to the energy density at the face of the transducer. This system worked reasonably satisfactorily at 1 mc/s. For lower frequencies the paddle tends to reflect some energy back towards the transducer. In order to overcome this defect Hall and Lamb[8.8] eliminated the paddle so that their liquid column and absorber formed one column. In order to obtain E they measured the admittance of the transducer when loaded by the liquid being studied and also when unloaded. In section 3.8 we showed that a plot of the real and imaginary components of a transducer's impedance gave a circle. One also obtains a circle by plotting the real and imaginary components of a transducer's input admittance. Using the terminology of chapter 3 the maximum conductance is given by $1/R_d + 4h^2 C_o^2/(Z_1 + Z_2)$ while the minimum conductance is given by $1/R_d$. The difference between these conductances equals the diameter of the admittance diagram circle, D. Hence we have:

$$D = \frac{4h^2 C_o^2}{Z_1 + Z_2}$$

If the transducer is unloaded, $Z_2 = 0$. In this case the circle diameter, D', is given by:

$$D' = \frac{4h^2C_o^2}{Z_1}$$

If a voltage V is applied across the transducer, the power W radiated into the liquid is given by:

$$W = \left[\frac{VZ_2}{Z_1 + Z_2}\right]^2 \frac{4h^2C_o^2}{Z_2} \quad . \quad . \quad . \quad . \quad \textbf{(8.16)}$$

Hence, if the radiating surface of the transducer has an area A, E is given from equations (8.14) and (8.16) as:

$$E = \frac{W}{Ac} = \frac{V^2(D' - D)D}{AcD'} \quad . \quad . \quad . \quad . \quad \textbf{(8.17)}$$

where c denotes the ultrasonic velocity. Hall and Lamb plotted the admittance circles from data obtained by means of an admittance bridge at frequencies around the transducer's resonant frequency. They estimate that their method gives an error of about 5% at 1560 kc/s rising to 11% at 130 kc/s.

8.7 The Interferometer

The interferometer[8.9] is used mainly to obtain accurate values for ultrasonic velocities at frequencies above 1 mc/s. A diagram of a typical instrument is given in Fig. 8.4. A transducer driven at its resonant frequency generates plane waves which travel through the medium being studied before being reflected back to the transducer. Thus the column terminated by the reflector acts like an acoustic transmission line so that the impedance seen by the transducer will be that given by equation (1.51). Since the transducer is operating at resonance this acoustic impedance will be transformed directly through the ideal transformer of the transducer equivalent circuit and appear as an electrical impedance in parallel with the transducer's static capacitance. If this static capacitance is tuned out by shunting it with an inductance to form a circuit resonating at the transducer resonant frequency and the transducer is fed via a resistor R from a constant amplitude source, the current flowing from the source will vary as the distance between the transducer and reflector is changed. If we denote the generator voltage by E, the acoustic impedances transformed through the transducer ideal transformer by Z_o' and Z_t' for the medium and reflector respectively and neglecting the absorption in the medium, the current i flowing from the generator is given by:

$$|i| = E\left[\frac{Z_o'^2 + Z_t'^2 \tan^2 \beta l}{Z_o'^2(R + Z_t')^2 + (RZ_t' + Z_o'^2)^2 \tan^2 \beta l}\right]^{\frac{1}{2}} \quad \textbf{(8.18)}$$

where l denotes the transducer–reflector spacing and we have neglected electrical and mechanical losses of the transducer. The current shows a maximum when $l/\lambda = (2n + 1)/4$ and a minimum when $l/\lambda = n/2$ where n is an integer. Thus the maxima are spaced at $\lambda/2$ apart, as are also the minima. Taking absorption into account the variation in current with reflector position is as shown in Fig. 8.5. The exact shape of the

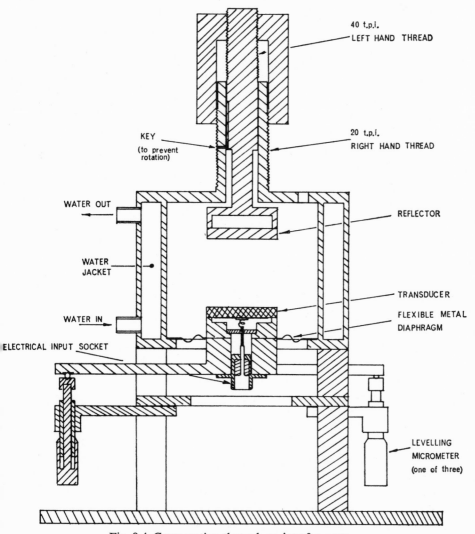

Fig. 8.4 Cross-section through an interferometer.

current variation with reflector position depends upon whether the medium is a gas or a liquid: with a gas the current minima are more sharply defined than the maxima while the opposite tends to hold for a liquid. This effect arises from the greater loading on the transducer produced by a liquid compared to a gas. If we denote the reflector–transducer separation at which the current has fallen to $1/\sqrt{2}$ of its peak value by l_1, equation (8.18) gives:

$$\tan \beta l_1 = \pm \frac{Z_0'}{Z_t'} \left[\frac{\left(\dfrac{R}{Z_t'} + 1 \right)^2}{\left[\dfrac{R}{Z_t'} + \left(\dfrac{Z_0'}{Z_t'} \right)^2 \right]} - 2 \right] \qquad . \quad . \quad \textbf{(8.19)}$$

When Z_0'/Z_t' is very small, that is when the medium is a gas, $\tan \beta l_1$ is also very small so that the narrow section of the current versus l curve is centred around a current minimum. As Z_0'/Z_t increases, as the medium is changed from a gas to a liquid, $\tan \beta l_1$ increases so that the curve around the current minimum broadens out and eventually the narrowest section appears around the current maximum.

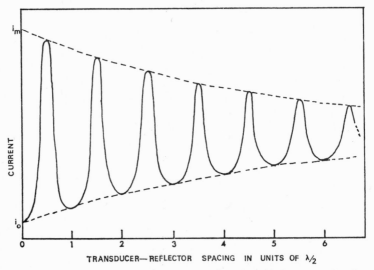

Fig. 8.5 Interferometer transducer current as a function of the separation between the transducer and the reflector for an absorbing liquid.

Since the interferometer theory assumes that the waves travelling in the medium are perfectly plane the transducer diameter must be many wavelengths across. Alternatively, the transducer and reflector can completely fill the cross-section of the cylinder containing the medium in which case the column acts like a waveguide in which many modes of propagation are possible[8.10]. Provided that the transducer acts like a piston source only the lowest or planewave mode should be propagated. The transducer and reflector should have accurately plane surfaces and be parallel, preferably to within a wavelength of light[8.11]. If the two surfaces are not parallel the current curve becomes assymetric, broader than it should be and in extreme cases secondary peaks can appear[8.12]. In order to set up an interferometer the current curves are plotted and the parallelism of the reflecting surfaces adjusted until the most symmetrical and sharpest curves are obtained. If measurements are being made by hand the electrical generator driving the transducer must exhibit a high degree of frequency stability so that the frequency does not shift appreciably over the long period of time necessary in order to plot the current maxima or minima accurately. Ideally a crystal controlled oscillator is called for. For gases of low absorption the position of a current minimum is difficult to fix exactly. An alternative method is to fix two positions, one on each side of the minimum, at which the current has the same value when the position of the minimum is given by the mean of the two positions[8.13]. In principle if one can measure the positions of the current minima or maxima over a distance of 1 cm and uses a 4-inch diameter drum micrometer reading to 4×10^{-5} cm the accuracy of velocity measurements should be about 1 part in 20,000. This accuracy has never yet been achieved mainly because of the breadth of the maxima or minima.

Since the envelope of the current depends upon the absorption it should be possible to obtain values for the absorption coefficient by interferometry as well as velocities. According to Mason[8.14] the absorption coefficient α can be found from the expression:

$$\tanh \alpha l = \sqrt{\frac{i_m}{i_{\max}} \frac{(i_m - i_{\max})(i_{\min} - i_o)}{(i_m - i_o)(i_m - i_{\min})}} \quad . \quad . \quad \textbf{(8.20)}$$

where i_m denotes the value of the maximum current extrapolated back to $l = 0$, i_o the corresponding value for the minimum current, and $i_{\max}$ and $i_{\min}$ the maximum and minimum currents at l.

An alternative type of interferometer employs a fixed reflector but variable frequency[8.15, 8.16]. As the frequency of the signal driving the transducer is varied, the transducer impedance will pass through a series of maxima and minima and provided that the transducer's resonant frequencies lie outside the frequency range being considered the condition that the transducer impedance has a maximum is that the distance between transducer and reflector is an integral number of half wavelengths. Thus if the frequencies of two adjacent impedance maxima are f_1 and f_2 and the transducer–reflector spacing is l, we have:

$$\frac{nc}{2f_1} = (n + 1)\frac{c}{2f_2} = l \quad . \quad . \quad . \quad . \quad \textbf{(8.21)}$$

where n is an integer.
Hence:

$$c = 2l(f_2 - f_1) \quad . \quad . \quad . \quad . \quad \textbf{(8.22)}$$

Equation (8.21) may in some circumstances be incorrect since the phase change upon reflection of the ultrasonic waves may not be zero or 180°. This may be taken into account by regarding the effective distance between transducer and reflector as being slightly greater than l.

F

The instruments so far described require a fairly large volume of liquid. McConnell and Mruk[8.17] have developed an interferometer requiring less than 0·1 cc of liquid. In their system a pulse of ultrasonic waves is passed into a thin film of the liquid contained between a wedge and a reflector. Multiple reflection of the pulse occurs within the liquid film and at each reflection at the liquid-wedge interface some of the pulse is transmitted back to the transducer where the reflected pulses are added together. As the reflector-wedge separation is altered the summed echoes at the transducer show minima with the separation changing by $\lambda/2$ between the minima. The wedge is used to separate desired from undesired echoes. Provided that the liquid's absorption is such that not fewer than 20 minima can be observed, McConnell and Mruk claim that they can achieve a velocity accuracy of 1 % with as little as 0·01 cc of liquid or 0·1 % with 0·1 cc.

As a method for finding velocities, interferometry is capable of providing results of high accuracy provided that the medium shows only moderate absorption so that many peaks can be found. It is less accurate for absorption. The disadvantages of the method are that it is tedious, the instrument must be made with a high degree of precision and must be very carefully set up.

8.8 Pulse Techniques

The pulse techniques originally developed for radar have been applied with considerable success to the measurement of ultrasonic absorptions and velocities[8.18]. A block diagram of a typical system is given in Fig. 8.6. A train of narrow pulses at a suitable pulse repetition frequency is generated by the pulse generator. Each pulse switches on the pulsed oscillator for the duration of the pulse. Thus a pulse modulated radio frequency voltage with a frequency equal to the transmitting transducer's resonant frequency is applied to the transducer which emits pulse modulated ultrasonic waves. These travel through the medium being studied until they hit the receiving transducer where they are converted into electrical signals, amplified by the amplifier and applied to the vertical deflection plates of the oscilloscope. Each pulse from the pulse generator also triggers off the triangular time base waveform which is applied to the horizontal deflection plates of the oscilloscope. The signal from the receiving transducer will be delayed relative to the start of the time base by the time required for the ultrasonic wave pulse to travel through the medium and thus it will appear on the face of the oscilloscope at a distance from the start of the time base dependent upon the separation of the transducers, the wave velocity and the time base speed. If we know the transducer separation and the time base speed we can find the ultrasonic velocity from the distance along the time base at which the received signal appears. This method, however, is not particularly accurate.

As the separation between the transducers is increased the signal from the receiving transducer will decrease because of the attenuation in the medium. Thus if we can measure the variation in pulse height on the face of the oscilloscope with transducer separation we can measure the absorption coefficient of the medium. A straightforward measurement using a graticule on the face of the oscilloscope will give only moderate accuracy and does not take into account a possible variation in amplifier gain with the amplitude of the signal. A better method is to use a comparison pulse, generated by another pulse generator and pulsed oscillator with the same radio frequency as that applied to the transmitting

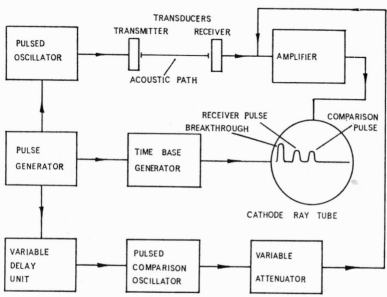

Fig. 8.6 Block diagram of a typical pulse system for ultrasonic absorption measurements.

transducer. This pulse is delayed on that from the main pulse generator by a variable time and applied to the amplifier through a variable attenuator. With this system the comparison pulse delay is adjusted until it and the signal from the receiver lie side by side and the comparison pulse amplitude is adjusted by means of the attenuator until its amplitude equals that of the received ultrasonic pulse. The separation between the transducers is then changed and the process repeated. By this means the amplitude of the received pulse can be measured to within a fraction of a decibel and any gain variation of the amplifier is automatically compensated for.

In order to measure the ultrasonic velocity the frequency of the comparison pulse is adjusted to exactly equal that of the main pulse. This can be achieved by arranging that both the main and comparison pulsed oscillators can be switched to run continuously so that their frequencies can be measured by means of standard frequency meters. Alternatively, the outputs from both oscillators can be mixed and the frequency of the comparison oscillator adjusted until zero beat frequency is obtained. The two oscillators are then switched back to pulse operation and the delay of the comparison pulse adjusted until it and the main received pulse overlap on the face of the oscilloscope. As the transducer separation is now altered the signal from the receiving transducer will vary in phase relative to the comparison pulse so that the amplitude of the combined main and comparison pulses will vary, being large when the two signals are in phase and small when in antiphase. The change in transducer separation between two in-phase conditions will equal one ultrasonic wavelength hence provided the ultrasonic frequency is known the velocity can be calculated[8.19].

Instead of the two transducers, one for transmission and one for reflection, only one, combining both functions, is sometimes used[8.20,8.21]. In this case the second transducer is replaced by a movable reflector. This system allows double the path length for the ultrasonic waves for the same volume of fluid compared with the two transducer system which is advantageous for low absorption media. Its disadvantage is that since the amplifier and pulsed oscillator are both across the same transducer, the amplifier is easily overloaded by the oscillator signal and may take some time to recover.

In highly absorbing media the ultrasonic path length may have to be kept small in order that the signal from the receiving transducer is not greatly affected by the noise from the amplifier. The signal to noise ratio cannot be increased by increasing the radiated intensity since the distorted ultrasonic waveform resulting from the finite effects of the second order terms in the wave equation will show a higher absorption than for the undistorted waveform because of the presence of harmonic components. With a path length of 1 cm in most liquids the time of travel of the ultrasonic pulse from one transducer to another is of the order of 10 microseconds or less. Since the amplifier will probably be paralysed by breakthrough from the transmitted pulse and also, since the pulse edges will be rounded, this time interval may be too short to give accurate reproduction of the received signal. The time interval can be lengthened by interposing fused quartz delay lines[8.18] between the transducers and the liquid as illustrated in Fig. 8.7. With this system multiple echoes will occur within the quartz rods due to reflections at the ends of the rods. However, the required signal is always the first one to appear on the oscilloscope after the breakthrough signal.

The pulse width should be long enough for transient effects at the leading edge to have died away but not so long that its breakthrough to the amplifier interferes with either the signal from the receiving transducer or the comparison pulse, whichever occurs first in time. The transient effects at the edges of the pulse arise from three causes[8.22]. First, the oscillations of the pulsed oscillator may take several cycles to build up to a constant amplitude, secondly, the amplifier response will modify the pulse, particularly if its bandwidth is insufficient and phase response markedly non-linear and, thirdly, the transducer bandwidth, or Q, will round the edges of the pulse. Ideally the transducer Q should

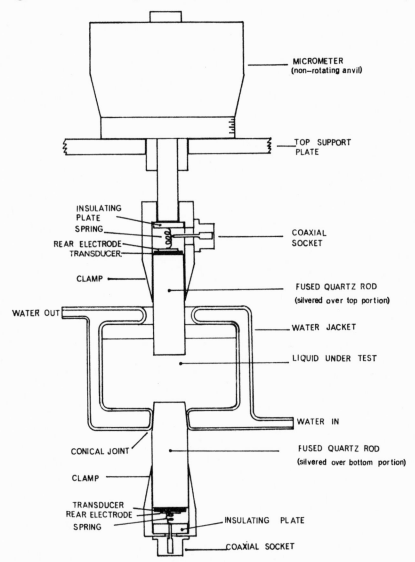

Fig. 8.7 Absorption cell for use with pulse techniques.

be as low as possible which implies that it should be heavily loaded acoustically. If the quartz delay lines are not used the transducers should be backed with tungsten loaded epoxy resin[8.23].

The choice of the pulse repetition frequency is governed by the time required for all the effects of one pulse to have died away. This frequency should be sufficiently low for the amplitude of all the multiply reflected signals between the two transducers to have become negligibly small but not so low that the oscilloscope picture flickers.

The liquid being studied should be thermostatted carefully so that temperature and hence density gradients do not exist since these scatter ultrasonic radiation and could lead to false values for the absorption coefficient. Another source of false values for the absorption coefficient arises from the fact that the wavefront from the transducer is not exactly plane—the necessary corrections have already been discussed in section 2.3c.

Finally, the attenuator should be adjustable in small increments. The most satisfactory type is the piston attenuator[8.24] since, provided that it is worked in its cut-off mode, the attenuation is an exponential function of the piston travel. This means that a plot of piston position against transducer separation will be a straight line from whose slope the absorption coefficient is easily calculated.

If only velocity measurements are required the sing-round techniques developed by Greenspan[8.25] will give accurate results. He uses two transducers situated a constant and accurately known distance apart. The transmitting transducer generates a pulse of ultrasonic waves which travels to the receiving transducer. When this pulse arrives it triggers off a pulsed oscillator which energises the transmitting transducer again. Thus the pulse repetition frequency of the pulsed oscillator will be exactly inversely proportional to the time required for the acoustic pulse to travel between the two transducers. Since this frequency can be measured to a high accuracy the acoustic velocity can be measured to a similar accuracy provided that the transducer separation is known.

References

8.1 Angona, F., *J. Acoust. Soc. Amer.*, **25**, 1111 (1953).
8.2 Edmonds, P. D. and Lamb, J., *Proc. Phys. Soc.*, **71**, 17 (1958).
8.3 Harlow, R. G., *Proc. 3rd Int. Cong. Acoust.*, p. 290 (1959).
8.4 Richardson, E. G., *Ultrasonic Physics*, p. 81. Elsevier (1962).
8.5 Karpovitch, J., *J. Acoust. Soc. Amer.*, **26**, 819 (1955).
8.6 Andrae, J. H. and Edmonds, P. D., *Proc. 3rd Int. Cong. Acoust.*, p. 556 (1959).
8.7 Piercy, J. E. and Lamb, J., *Proc. Roy. Soc.*, A, **226**, 43 (1954).
8.8 Hall, D. N. and Lamb, J., *Proc. Phys. Soc.*, B, **73**, 354 (1959).
8.9 Hubbard, J. C., *Phys. Rev.*, **46**, 525 (1934).
8.10 Stewart, J. L. and Stewart, E. S., *J. Acoust. Soc. Amer.*, **24**, 22 (1952).
8.11 Hubbard, J. C., *Amer. J. Phys.*, **8**, 207 (1940).
8.12 Alleman, R. S., *J. Acoust. Soc. Amer.*, **13**, 23 (1941).
8.13 Sette, D., Basula, A. and Hubbard, J. C., *J. Chem. Phys.*, **23**, 787 (1955).
8.14 Mason, W. P., *Piezoelectric Crystals and Their Application to Ultrasonics*, p. 319. Van Nostrand (1956).
8.15 Ballon, I. W. and Hubbard, J. C., *J. Acoust. Soc. Amer.*, **16**, 101 (1944).

8.16 Borgnis, F. E., *J. Acoust. Soc. Amer.*, **24**, 19 (1952).

8.17 McConnell, R. A. and Mruk, W. F., *J. Acoust. Soc. Amer.*, **27**, 672 (1955).

8.18 Andrae, J. H., Bass, R., Heasell, E. L. and Lamb, J., *Acustica*, **8**, 3 (1958).

8.19 Mason, W. P., Baker, W. O., McSkimin, H. J. and Heiss, J. H., *Phys. Rev.*, **73**, 1074 (1948).

8.20 Pinkerton, J. M. M., *Proc. Phys. Soc.*, B, **62**, 162, 186 (1949).

8.21 Rapuano, R. A., *Phys. Rev.*, **72**, 78 (1947).

8.22 Gooberman, G., *Ultrasonic Techniques in Biology and Medicine*, ed. Brown, Chapter 4. Iliffe (1967).

8.23 Lutsch, A., *J. Acoust. Soc. Amer.*, **34**, 131 (1962).

8.24 Fink, D. G., *Radar Engineering*, p. 601. McGraw-Hill (1947).

8.25 Greenspan, M. and Tschiegg, C. E., *J. Res. Natl. Bur. Standards*, **59**, 249 (1957).

9
Shear Waves

9.1 Introduction

In the preceding chapters we have been almost exclusively concerned with the properties of longitudinal waves. However, transverse or shear waves can be propagated in solids and to a certain extent in liquids. As far as solids are concerned, shear waves behave very similar to longitudinal waves being propagated with a velocity given by equation (1.21). Liquids are more interesting since they are not normally regarded as possessing a shear rigidity and therefore we would not expect them to sustain shear waves. However, liquids do show viscosity which, as we shall show, does permit them to sustain shear waves although their attenuation with distance is extremely rapid. In addition experimental results from certain liquids, in particular low molecular weight polymers and polymer solutions, require the postulation of a relaxing shear rigidity for their explanation.

9.2 Propagation of Shear Waves in Liquids

The equation of motion for shear waves in any medium was derived in section 1.4 as equation (1.19), viz.:

$$\frac{\partial^2 \xi_x}{\partial t^2} = \left(\frac{\mu}{\rho_o}\right)\frac{\partial^2 \xi_x}{\partial y^2} \qquad . \quad . \quad . \quad . \quad \textbf{(9.1)}$$

which describes a wave travelling in the y direction with its particle motion in the x direction. According to equation (6.5), the shear rigidity μ is related to the fluid viscosity η for sinusoidal motion by:

$$\mu = j\omega\eta \qquad . \quad . \quad . \quad . \quad . \quad \textbf{(9.2)}$$

Also, since the particle velocity $u_x = j\omega\xi_x$, we can combine equations (9.1) and (9.2) to give:

$$\frac{\partial u_x}{\partial t} = \left(\frac{\eta}{\rho_o}\right)\frac{\partial^2 u_x}{\partial y^2} \qquad . \quad . \quad . \quad \textbf{(9.3)}$$

The solution to equation (9.3) can be written as:

$$u_x = U_o' e^{j\omega t}$$

where U_o' may be complex. Substituting into equation (9.3) gives:

$$j\omega U_o' = \left(\frac{\eta}{\rho_o}\right)\frac{\partial^2 U_o'}{\partial y^2} \qquad \ldots \quad \textbf{(9.4)}$$

If we now let $U_o' = U_o e^{ay}$ and substitute into equation (9.4) we find:

$$a = \pm\ \sqrt{j\omega\rho_o/\eta}$$

We have the identity:

$$\sqrt{j} = \sqrt{\tfrac{1}{2}}(1+j)$$

hence:

$$a = \sqrt{\frac{\pi f \rho_o}{\eta}}(1+j) \quad . \quad \ldots \quad \textbf{(9.5)}$$

Thus we have:

$$u_x = U_o e^{j\omega t} e^{-\sqrt{\frac{\pi f \rho_o}{\eta}}(1+j)y} \quad . \quad . \quad \textbf{(9.6)}$$

According to equation (9.6) the absorption coefficient for shear waves is given by:

$$\alpha_{\text{shear}} = \sqrt{\frac{\pi f \rho_o}{\eta}} \qquad . \quad . \quad . \quad \textbf{(9.7)}$$

For water the value of the absorption coefficient at a frequency of 10 kc/s is $1\cdot78\ 10^3$ nepers/cm. This is extremely high, compared, for example, with the absorption coefficient at the same frequency for longitudinal efficient is 10^{-6} nepers/cm. Thus although shear waves can exist in a liquid they are rapidly attenuated and penetrate only a very short distance into the medium.

9.3 Specific Acoustic Impedance for Shear Waves in a Liquid

The specific acoustic impedance for shear waves is defined in an analogous manner as for longitudinal waves as the ratio of shear stress to particle velocity. The shear stress is given by:

$$\text{shear stress} = -\eta\frac{\partial u_x}{\partial y} \qquad . \quad . \quad . \quad \textbf{(9.8)}$$

If we now combine equations (9.6) and (9.8) with the definition of impedance we find:

$$\text{specific acoustic impedance } Z_{\text{shear}} = \sqrt{\pi f \rho_o \eta}(1+j) \qquad \textbf{(9.9)}$$

The specific acoustic impedance for shear waves differs from that for longitudinal waves in that it is a function of frequency and is complex, with the magnitudes of the real and imaginary components being equal.

We shall be describing later how the shear acoustic impedances of liquids are measured. For 'normal' liquids the resistive and reactive components of the impedances are found to be equal but this equality does not hold for polymeric liquids and solutions for which the resistive component is found to be greater than the reactive one. We can interpret this divergence by assuming that the liquid possesses a shear rigidity as well as a viscosity.

If we take equation (9.9) and combine it with equation (9.2) we obtain:

$$Z_{\text{shear}} = \sqrt{\rho_0 \mu} \qquad \cdots \qquad \textbf{(9.10)}$$

We now assume that μ is complex with the real part representing a shear rigidity and the imaginary part representing, as before, a viscosity. We shall also now modify our terminology to bring it into line with that normally employed for viscoelastic effects, and we write:

$$Z^2_{\text{shear}} = (R + jX)^2 = \rho_0 \mu = \rho_0 G^* = \rho_0(G' + jG'') \quad \textbf{(9.11)}$$

where $\mu = G^*$, the complex shear modulus for the liquid, with components G' and G''.

If we expand $(R + jX)^2$ in equation (9.11) and equate real and imaginary terms we find:

$$R^2 - X^2 = \rho_0 G' \qquad \cdots \qquad \textbf{(9.12)}$$

$$2RX = \rho_0 G'' \qquad \cdots \qquad \textbf{(9.13)}$$

From equation (9.12) we see that if R exceeds X the liquid possesses a shear rigidity G'. The imaginary term G'' simply represents the liquid's viscosity since $G'' = \omega\eta$.

9.4 Transmission Line Analogy for Viscoelastic Effects

Electrical transmission lines furnish a useful analogy in the study of shear-wave propagation in the same manner as for longitudinal waves. In Fig. 1.7 we represented an electrical transmission line by a series connection of inductances and a shunt connection of capacitances. For shear waves we find it convenient to generalise this circuit to the form shown in Fig. 9.1. It is shown in standard works on transmission lines[9.1]

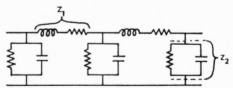

Fig. 9.1 General equivalent circuit for an electrical transmission line with losses.

that the characteristic impedance of such a line is given by $\sqrt{Z_1 Z_2}$ and the attenuation and phase constant by the equation:

$$\sqrt{Z_1/Z_2} = \alpha + j\beta \qquad \cdots \qquad \textbf{(9.14)}$$

If we consider a liquid for which $G' = 0$, equation (9.9) shows that its impedance is complex from which we infer that the j operators in the impedances Z_1 and Z_2 do not cancel in the expression Z_1Z_2 as they do for longitudinal waves. In fact we find that for shear waves the appropriate analogies are:

$$Z_1 = j\omega L = j\omega\rho_o \text{ (as for longitudinal waves)}$$
$$Z_2 = r = \eta$$

In this case, illustrated in Fig. 9.2, we have, therefore, that:

$$Z_{\text{shear}} = \sqrt{j\omega\rho_o\eta} = \sqrt{\pi f\rho_o\eta}\,(1 + j) \quad . \quad . \quad \textbf{(9.15)}$$

Equation (9.15) is identical to equation (9.9).

We now have to modify Fig. 9.2 to take into account the presence of

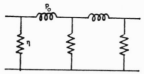

Fig. 9.2 Transmission line analogue for shear waves in a purely viscous medium.

rigidity. If the medium through which the shear waves were travelling were a solid the transmission line analogy would contain Z_1 as before but Z_2 would now be capacitive as for longitudinal waves. With a viscoelastic medium we would expect the electrical analogue to be intermediate between that for a normal or Newtonian liquid and a solid and, in practice, we find that we can represent the viscoelastic medium with, as before, Z_1 being inductive but Z_2 as a parallel combination of resistance, representing a viscosity, and capacitance, representing a rigidity, as illustrated in Fig. 9.3. The reader should note that the rigidity μ of

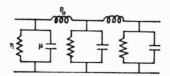

Fig. 9.3 Transmission line analogue for shear waves in a liquid possessing both viscosity and rigidity.

Fig. 9.3 represents an 'internal' constant rigidity of the liquid whereas G' represents the measured effective rigidity of the liquid which is not constant but is a function of frequency.

Assuming the correctness of our transmission line analogue we can now proceed to derive expressions for the resistive and reactive components of both the specific acoustic impedance and shear modulus. After some manipulation we find:

$$R^2 = \frac{\rho_o\mu}{2}\frac{\omega^2\tau^2}{1 + \omega^2\tau^2}\left[1 + \sqrt{1 + \frac{1}{\omega^2\tau^2}}\right] \quad . \quad . \quad \textbf{(9.16)}$$

$$X^2 = \frac{\rho_0 \mu}{2} \frac{\omega^2 \tau^2}{1 + \omega^2 \tau^2} \left[\sqrt{1 + \frac{1}{\omega^2 \tau^2}} - 1 \right] \quad . \quad . \quad (9.17)$$

$$G' = \mu \frac{\omega^2 \tau^2}{1 + \omega^2 \tau^2} \quad . \quad . \quad . \quad . \quad . \quad . \quad . \quad (9.18)$$

$$G'' = \mu \frac{\omega \tau}{1 + \omega^2 \tau^2} \quad . \quad . \quad . \quad . \quad . \quad . \quad . \quad (9.19)$$

where $\tau = \eta/\mu$ and as it has the dimension of time we can regard it as a relaxation time. The frequency dependance of R, X, G' and G'' are illustrated in Fig. 9.4 and 9.5. The value of the relaxation time can be

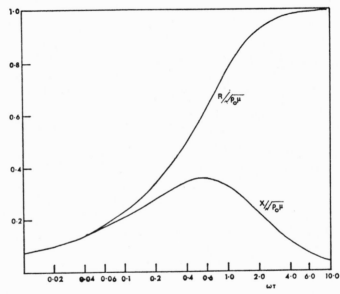

Fig. 9.4 Normalised values for the real and imaginary parts of the complex shear specific acoustic impedance of a liquid as a function of frequency and relaxation time.

found from plots of either G' or G'' since $\omega\tau = 1$ when G' has half its asymptotic high-frequency value or when G'' has a maximum.

As an example of the value of shear wave studies we will consider the well-known results of W. P. Mason[9.2] and his co-workers for a 1 % solution of polyisobutylene in cyclohexane which are reproduced in Fig. 9.6. The rigidity G' of the solution in this case does not follow equation (9.18) which Mason interprets as being due to the presence of more than one relaxation associated with the motion of the dissolved macromolecules. Good agreement between theoretical and experimental data was obtained by postulating three relaxation mechanisms arising from the macromolecular movement coupled with the effect of the solvent viscosity. Since these effects are additive they can be represented by the transmission line shunt arm shown in Fig. 9.7. Mason suggests that η_1

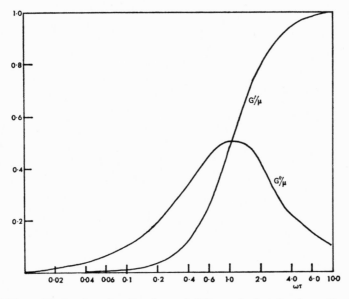

Fig. 9.5 Normalised values for the shear rigidity and viscosity for a liquid as a function of frequency and relaxation time.

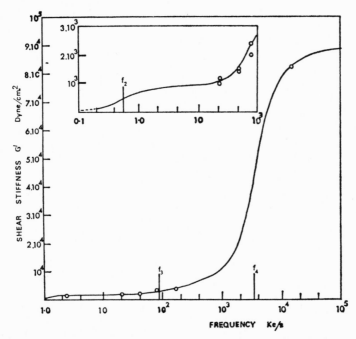

Fig. 9.6 Shear stiffness as a function of frequency for a 1% solution of polyiso-butylene in cyclohexane. (After Mason *et al.* (9.2).)

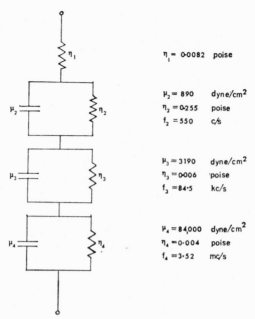

$\eta_1 = 0.0082 \quad$ poise

$\mu_2 = 890 \quad$ dyne/cm^2
$\eta_2 = 0.255 \quad$ poise
$f_2 = 550 \quad$ c/s

$\mu_3 = 3190 \quad$ dyne/cm^2
$\eta_3 = 0.006 \quad$ poise
$f_3 = 84.5 \quad$ kc/s

$\mu_4 = 84,000 \quad$ dyne/cm^2
$\eta_4 = 0.004 \quad$ poise
$f_4 = 3.52 \quad$ mc/s

Fig. 9.7 Shunt arm of transmission line analogue required to account for the data of
Fig. 9.6.

is the viscosity of the pure solvent, μ_2 the configurational rigidity of the
macromolecule arising from changes in the distance between its ends
while η_2 represents the viscous drag of the solvent flowing through the
molecule as the latter attempts to change its dimensions. μ_3 may represent temporary entanglements of one part of the molecule with another
part and the associated viscosity η_3 presumably represents slipping at the
points of entanglement. μ_4 may represent a rigidity arising from hin-
drances to rotation of one segment of the polymer chain relative to
another and η_4 its associated viscosity.

9.5 Shear Wave Techniques

In order to find the components of the complex shear modulus for a
liquid we have to measure its complex specific acoustic impedance. A
variety of instruments have been developed for this purpose based either
on measuring the reaction of the medium on a shear wave generator or
on the phase and amplitude changes occurring at the reflection of a
shear wave at an interface between a solid non-viscoelastic medium and
the medium under test. We shall consider one example of each type of
instrument.

9.5a The torsional crystal

If a cylinder of quartz with its axis parallel to the x axis of the parent
crystal is fitted with electrodes as shown in Fig. 9.8 a face shear in the xy

plane will be set up when the electrodes are energised which reverses its direction as one travels along a diameter in the z direction so that effectively the outer curved surface of the cylinder rotates and sets up a torsional mode of vibration when the electrodes are energised by a voltage at a suitable frequency[9.3].

The frequency dependance of the electrical impedance of the torsional crystal will be similar to that for an X-cut crystal, becoming resistive at the resonant frequency if the static electrode capacitance is tuned out. The value of the resistance will depend upon the acoustic loading at the cylindrical surface occasioned by the shear waves radiating into the medium enveloping the crystal. However, the acoustic impedance for shear waves is complex and while the resistive component

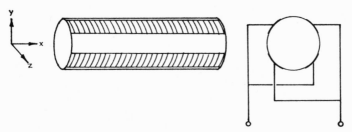

Fig. 9.8 Torsional piezoelectric transducer.

is simply transformed into an equivalent electrical resistance at the crystal terminals the reactive term will reduce the crystal's resonant frequency below that when it is operated in a vacuum. If we denote the change in electrical resistance at resonance due to loading by the medium by ΔR_e and the change in resonant frequency by Δf, Mason[9.3] has shown that:

$$R = \Delta R_e / K_1$$
$$X = -\Delta f / K_2$$

where R and X refer to real and imaginary components of acoustic impedance of the medium. While it is possible to compute values for the crystal constants K_1 and K_2 these are better found by calibrating with Newtonian liquids of known viscosities.

In operation, the torsional crystal is suspended by four wires soldered to its electrodes. Electrically the crystal forms one arm of a bridge, two of the other arms contain resistances and the final arm contains a capacitance which is used to balance out the static capacitance of the crystal. The value of this capacitance is best found by measuring the effective crystal capacitance at a frequency much higher than the crystal's resonant frequency[9.4]. With the crystal mounted in a vacuum the bridge is balanced and the crystal is then immersed in the liquid under test and the bridge again balanced by varying the drive frequency and adjusting the resistive arms of the bridge to give Δf and ΔR_e. This system will give accuracies of the order of 1 % or better over a frequency range of 10 to 150 kc/s.

If the liquid's viscosity is greater than about 10 poise the electrical resistance change becomes difficult to measure since the value of the resistance becomes so large that it is effectively shunted by the much smaller reactance of the crystal capacitance. One solution is to use a different crystal such as ammonium dihydrogen phosphate which, for the same acoustic loading, presents a much lower resistance to capacitance ratio than quartz. However, this material is more fragile than quartz and presents difficulties in making connections to the electrodes. An alternative technique measures the change in phase and amplitude of a torsional wave in a rod caused by the surrounding liquid being studied and is described in the next section.

9.5b Reflection method

In this method a torsional crystal is glued to one end of a cylindrical rod of glass or nickel iron. A pulsed radio frequency voltage at the crystal's resonant frequency is applied to the crystal which emits pulsed torsional waves. These travel down the rod, are reflected at the far end and then return to the crystal where they are converted back into electrical signals. According to Mason and McSkimin[9.4] the specific acoustic impedance Z_{shear} of the liquid surrounding the rod is given by:

$$Z_{shear} = \left(\frac{\rho c}{4}\right)\frac{a}{l}[\Delta A + j\Delta B] \quad . \quad . \quad . \quad \textbf{(9.20)}$$

where ρ and c denote the density and wave velocity for the rod of radius a of which a length l is covered by the liquid. ΔA and ΔB denote the change in attenuation and phase respectively between when the rod is surrounded by liquid and in air. The electrical circuitry associated with this method is given in Fig. 9.9. The output from a continuously running oscillator at the crystal's resonant frequency is applied through a phase-shifting network to a gate which interrupts the oscillator signal to apply a pulsed waveform to the crystal. The crystal itself is connected into a bridge which is balanced until the signal applied to the transducer does not appear at the input to the buffer amplifier A_2 although the output from the crystal of the received pulse is passed with little loss directly into A_2. By this means A_2 is not overloaded by the initial pulse applied to the crystal. The output from the oscillator is fed through an attenuator and buffer amplifier A_1 whose output is combined with that from A_2 and fed to the Y plates of an oscilloscope. With the rod in air the phase shifter and attenuator are adjusted until the first received pulse and the signal passing through A_1 balance out as indicated by the trace on the oscilloscope. The rod is now immersed in the liquid under test and the phase and attenuation again adjusted until cancellation is obtained. The change in settings of the phase shifter and attenuator give ΔB and ΔA respectively. This method has been found suitable for measuring liquids with viscosities ranging from 10 to 1000 poise over a frequency range of from 20 to 200 kc/s with an accuracy of about 10%.

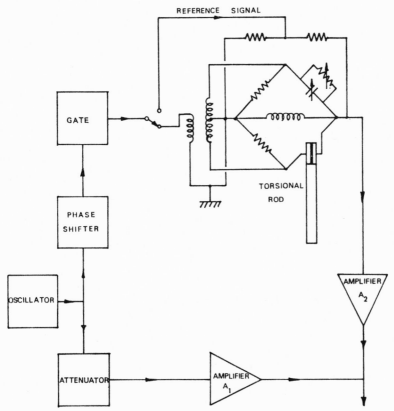

Fig. 9.9 Block diagram of system using torsional transducers. (After Mason *et al.* (9.4).)

Since the resonant frequency of a torsional crystal is inversely proportional to its length, the crystal becomes too small to be practicable at frequencies greater than 500 kc/s and at higher frequencies an alternative method has to be used.

The method[9.5] used at frequencies ranging from 3 mc/s to 100 mc/s uses the amplitude and phase change accompanying the reflection of a shear wave at the interface between a solid and the liquid being studied. The amplitude change arises from the difference in acoustic impedances between the solid and liquid while the phase change arises from the fact that while the impedance of the solid is resistive that of the liquid is complex. The system developed by Mason and his co-workers is shown in Fig. 9.10. A *Y*- or *AT*-cut quartz crystal generating shear waves is attached to one end of a quartz rod. The shear waves, with their particle motion parallel to the reflecting surface, so that only shear waves are reflected, are reflected from the top surface of the rod and travel to a second crystal where they are converted back into an electrical signal. Two such systems are driven from the same pulsed oscillator via phase shifting and attenuating networks but the liquid under investigation is

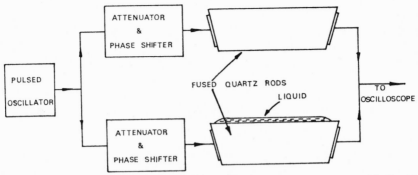

Fig. 9.10 Reflection technique for making shear-wave measurements with liquids. (After Mason *et al.* (9.5).)

spread on the reflecting surface of one rod only. The outputs from the two receiving crystals are combined and fed to the Y plates of an oscilloscope. With no liquid present the phase shifting and attenuating networks are adjusted until the outputs from the two receiving crystals cancel. Liquid is now spread on one reflecting surface and the change in attenuation ΔA and phase ΔB required to rebalance the circuit measured. The shear acoustic impedance of the liquid is given by[9.6]:

$$Z_{\text{shear}} = Z_q \cos \varphi \left[\frac{1 - (\Delta A)^2 + 2j(\Delta A) \sin (\Delta B)}{1 + (\Delta A)^2 + 2(\Delta A) \cos (\Delta B)} \right] . \quad \textbf{(9.21)}$$

where Z_q denotes the shear impedance of quartz and φ the angle between the wave normal and the reflecting surface. For maximum sensitivity φ should be large and in practice is usually made 80°. This method has the advantage that only small quantities of liquid are required since provided that the liquid film is thicker than about 0·02 mm it may be regarded as being infinitely thick on account of the very high absorption of shear waves.

9.6 Reduced Variables

We showed in section 7.4 that the effective frequency range over which relaxation data for gases could be obtained can be greatly extended by varying the pressure as well as the frequency. A similar situation holds for viscoelastic investigations with temperature instead of pressure as another variable to supplement frequency. The object is to 'reduce' the values for the moduli and viscosities obtained at various temperatures to the values appertaining to some reference temperature[9.7].

The moduli G' and G'' are assumed to have the same temperature dependence as that derived from the theory of rubberlike elasticity for polymer solutions, viz.:

$$G = \left(\frac{C_w}{100} \right) \left(\frac{R}{M} \right) T\rho_s . \quad . \quad . \quad . \quad . \quad \textbf{(9.22)}$$

Where C_w denotes the concentration of the polymer in weight per cent, R the gas constant, M the molecular weight, T the absolute temperature and ρ_s the solution density. The reduced modululs G_{red} will be the value of G at some reference temperature T_o at which the solution has a density ρ_o. We have, therefore, from equation (9.22) that:

$$G_{red} = G\frac{T_o\rho_o}{T\rho} \qquad \ldots \quad \ldots \quad (9.23)$$

where ρ denotes the solution density at temperature T. Equation (9.23) is assumed to be valid for both G' and G''.

The viscosities associated with viscoelasticity are assumed to depend upon the solvent viscosity so that the reduced viscosity η_{red} can be derived from the measured value η_1 if the solvent viscosities η_T and η_o at two temperatures T and T_o are known by means of the relation:

$$\eta_{red} = \eta_1\frac{\eta_o}{\eta_T} \qquad \ldots \quad \ldots \quad (9.24)$$

From equations (9.23) and (9.24) we can derive the value for the relaxation time τ_o at T_o as follows:

$$\tau_o = \frac{\eta_{red}}{G_{red}} = \frac{\eta_1}{G} \cdot \frac{\eta_o}{\eta_T} \frac{T\rho}{T_o\rho_o} \qquad \ldots \quad \ldots \quad (9.25)$$

that is:

$$\tau_o = \tau_{exp.}/\alpha_T \qquad \ldots \quad \ldots \quad \ldots \quad (9.26)$$

where:

$$\alpha_T = \eta_T T_o\rho_o/\eta_o T\rho \qquad \ldots \quad \ldots \quad \ldots \quad (9.27)$$

To use the method of reduced variables the moduli and viscosities are reduced to those for the reference temperature by means of equations (9.23) and (9.24) and at the same time the frequencies are multiplied by α_T calculated from equation (9.27). With low viscosity solvents the available temperature range gives α_T a range of only two or three but with higher viscosity solvents a range of several decades can sometimes be achieved.

While the method of reduced variables is extremely useful it is strictly valid only for very dilute polymer solutions. Modifications have been proposed by De Mallie[9.8] and his co-workers to deal with more concentrated solutions but their work is beyond the scope of this book and the reader is referred to their paper for further details.

One final point which should be noted is that throughout this chapter we have assumed that a solution displays only one or a small number of discrete relaxation frequencies. Since polymers are rarely monodisperse this assumption may not always be true and one may have to contend with a distribution of relaxation times when the analysis of experimental results becomes considerably more difficult[9.9].

References

9.1 Koehler, G., *Circuits and Networks*, p. 221. Macmillan (1955).

9.2 Baker, W. O., Mason, W. P. and Heiss, J. H., *J. Poly. Sci.*, **8**, 129 (1952).

9.3 Mason, W. P., *Trans. Amer. Soc. Mech. Eng.*, **69**, 359 (1947).

9.4 Mason, W. P. and McSkimin, H. J., *Bell System Tech. J.*, **31**, 122 (1952).

9.5 Mason, W. P., Baker, W. O., McSkimin, H. J. and Heiss, J. H., *Phys. Rev.*, **75**, 936 (1949).

9.6 O'Neil, H. T., *Phys. Rev.*, **75**, 928 (1949).

9.7 Ferry, J. D., *J. Amer. Chem. Soc.*, **72**, 3746 (1950).

9.8 De Mallie, R. B., Birnboim, M. H., Frederick, J. E., Tschoegl, N. W. and Ferry, J. D., *J. Phys. Chem.*, **66**, 536 (1962).

9.9 Barlow, A. J. and Lamb, J., *Proc. 3rd Int. Cong. Acoust.*, p. 565 (1959).

10

Propagation in Solids

10.1 Introduction

The velocity of an ultrasonic wave in a solid medium is related to the elastic coefficients of the medium while the ultrasonic absorption gives information about the medium's internal structure. We shall treat the various absorption mechanisms so far discovered for solids somewhat briefly with the exception of phonon-electron and phonon-spin interactions in semiconductors since while the former are well documented in many books the latter is still available only in journals and, at this moment, shows promise of becoming a very important development.

10.2 Adiabatic and Isothermal Elastic Constants

If we can measure the velocity of ultrasonic waves in a solid we can derive the solid's elastic constants from equations (1.20) and (1.21). The derived constants, however, may, in principle, be either the adiabatic or isothermal values depending upon the frequency at which they are measured. So far, in this book, we have tacitly assumed throughout that ultrasonic wave propagation is an adiabatic process whereas it can become isothermal at high frequencies. According to equation (6.16) the rate of loss of heat due to thermal conductivity is proportional to $-\dfrac{\partial^2 T}{\partial x^2}$. If we now consider equation (6.17) and neglect absorption we have:

$$-\frac{\partial^2 T}{\partial x^2} = \frac{4\pi^2}{\lambda^2}(T - T_a)$$

Thus while the time in which heat can be transferred from one part of the wave to another decreases as the frequency increases this does not compensate the frequency squared term governing the rate of heat flow. Thus a wave becomes less adiabatic as its frequency is increased and while this effect holds for waves in all media it will only become of practical importance in good thermal conductors. According to Mason[10.1] the frequency around which the transformation from adiabatic to isothermal conditions occurs is given approximately by:

$$f \approx \frac{\rho_0 C_v c_0{}^2}{2\pi K_h}$$

169

For tin, Mason calculates the frequency to be $4 \cdot 3 \ 10^4$ mc/s. This frequency is only just within the bounds of experimental possibility but if the temperature is reduced to $4°K$ the critical frequency falls to $0 \cdot 14$ mc/s. This effect has not yet been confirmed. Stern[10.2] and his co-workers found no appreciable change in the velocity in sodium at $0 \cdot 26$ mc/s over a temperature range of from $2°K$ to $78°K$.

10.3 Absorption in Solids

The causes of absorption in solids can be roughly divided into three groups; thermal effects, imperfections and electron-phonon effects. As with liquids and gases the existence of thermal conductivity leads to absorption with an absorption coefficient given by [10.3]:

$$\alpha = \frac{\omega^2}{2\rho_0 c_0^2}\left[\chi + 2\eta + \left(\frac{\lambda^\sigma - \lambda^\theta}{\lambda^\sigma + 2\mu}\right)\frac{K_h}{C_v}\right] \qquad . \quad . \quad (10.1)$$

where χ and η denote the compressional and shear viscosities respectively and λ^σ and λ^θ the adiabatic and isothermal Lamé constants respectively. This expression (10.1) is similar in form to that derived for liquids and gases (6.20).

Many solid media, particularly metals, contain anisotropic grains whose presence gives rise to ultrasonic absorption through either thermal or scattering effects. The thermoelastic or Zener effect[10.4] arises from the fact that differing grains present differing orientations to a longitudinal wave. Since the grains are crystalline their elastic properties depend upon the direction in which they are stressed and hence differing grains experience differing strains during the passage of a longitudinal wave. Therefore differing grains will experience differing temperature changes during the compression and dilatation caused by the ultrasonic wave so that a random temperature distribution will appear throughout the bulk of the medium. Rapid heat exchanges will take place between adjacent grains and this heat interchange will lead to an ultrasonic absorption with an absorption coefficient given by:

$$\alpha = \frac{\beta_h T}{2\pi C_v}\frac{\omega^2/\omega_h}{1 + (\omega/\omega_h)^2} \qquad . \quad . \quad . \quad (10.2)$$

where β_h denotes the linear thermal expansion coefficient of the medium at a temperature T and ω_h is defined by:

$$\omega_h \approx 0 \cdot 8 K_h/C_v L_c^2 \qquad . \quad . \quad . \quad (10.3)$$

where L_c denotes the diameter of the grains. The relaxation frequency for this effect is usually below 100 kc/s and the contribution to the total absorption will generally be small because of the wide variations in grain size normally present in any solid.

The presence of grains also leads to scattering of the incident radiation. Differing grains present differing elasticities to the ultrasonic waves and since the grain densities will all be equal acoustic mismatches will occur at grain boundaries and give rise to scattering leading to an absorption coefficient given by[10.5]:

$$\alpha = \frac{1}{2L_c}\left(\frac{Q_s}{A}\right). \qquad \cdots \qquad \textbf{(10.4)}$$

where Q_s/A denotes the ratio of the scattering cross-section of a grain to its actual cross-sectional area (cf. section 2.5a). Assuming the grains to be spherical leads to a frequency dependence of the absorption coefficient similar to the frequency variation in scattering cross-section shown in Fig. 2.10. This effect has been confirmed by Mason and McSkimin[10.6] with aluminium and their results are reproduced in Fig. 10.1. At low frequencies where the wavelength is much larger than

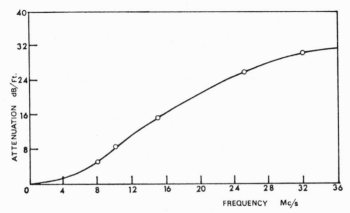

Fig. 10.1 Frequency dependence of the absorption of ultrasonic waves in aluminium.
(After Mason *et al.* (10.6).)

the grain size the absorption follows Rayleigh's fourth power scattering law but as the wavelength decreases the variation in absorption with frequency approaches a square law over the region where the grain size lies between 0·33 and 0·7 of a wavelength after which the absorption rises even more slowly and would appear to tend towards a constant value. Over the region in which the wavelength is of the same order as the grain size Huntingdon[10.7] has produced evidence that the absorption coefficient is directly proportional to the square of the product of the grain size and the frequency. At high frequencies where the wavelength is much smaller than the grain size the absorption coefficient is a linear function of frequency with the rate of increase in absorption with frequency being an inverse function of grain size. At these frequencies the propagation of the ultrasonic waves can be likened to a diffusion process with the scattered energy bouncing to and fro between the grains. The scattering will depend upon the number of collisions the wave makes in travelling a given distance and the distance between collisions will be of the order of the grain size. As this decreases the number of collisions and therefore the rate of energy loss increases.

Other forms of absorption due to grain size occur in ferromagnetic materials in which the grains are now the magnetic domains which change their wall positions under stress. Since the change in wall position requires a finite time to take place and involves an energy change an acoustic absorption will occur. The reader is referred to Mason's[10.8] and Bozorth's[10.9] books for further details.

A further cause of absorption in some media arises from the presence of interstitial and substitutional atoms. An interstitial atom is a foreign atom situated within the crystal lattice and a substitutional atom is a foreign atom occupying a position in the lattice normally occupied by a main lattice atom. Nitrogen atoms in the iron lattice form an example of interstitial atoms while zinc in copper, forming brass, is an example of a substitutional atom. In both cases a deformation of the lattice by means of a stress alters the preferred positions occupied by the foreign atoms. Since differing preferred positions possess differing energies a finite time is required for the atoms to move from one preferred position to another. If the applied stress is sinusoidal as is the case if an ultrasonic wave is present the time required for the foreign atoms to move will lead to a relaxation and hence absorption. Interstitial effects have been investigated by Snoek,[10.10] Dykstra and Ké[10.11] and substitutional effects by Zener[10.12].

10.4 Dislocations

Many solids for which the grains are negligibly small or non-existent, such as complete crystals, still show an ultrasonic absorption considerably greater than can be accounted for by thermal conductivity effects.

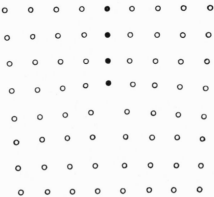

Fig. 10.2 Pictorial representation of one type of dislocation.

While the causes for this absorption are not yet completely understood several mechanisms have been postulated. If we restrict our discussion for the moment to insulating crystals one postulated cause of absorption arises from the presence of dislocations.

The measured mechanical strengths of crystals are usually several orders of magnitude smaller than the theoretical values and this discrepancy is attributed to the presence of imperfections in the crystal lattice known as dislocations. For example, a line or a plane of atoms may be displaced from its correct position in the lattice as shown in Fig. 10.2. It is assumed that this dislocation can either move through the lattice or vibrate under the influence of the ultrasonic wave and that this motion is damped in some way to lead to an energy loss. One suggested damping mechanism is phonon viscosity. Phonons are quantised thermal vibrations of the lattice and since ultrasonic waves are mechanical vibrations we may also regard these as being composed of low energy phonons. It can be shown[10.13] that we can attribute a viscosity to phonons equal to $E_oK_h/C_v\bar{c}^2$ where E_o denotes thermal energy per unit volume of the solid, K_h the thermal conductivity, C_v the specific heat per unit volume and $\bar{c}$ the average sound velocity for thermal phonons. This viscosity appears to provide a drag on the motion of the dislocations.

10.5 Phonon–Phonon Interactions

The absorption of ultrasonic waves in silicon crystals with dislocation densities varying by a factor of fifty to one between different crystals remains unchanged so that it appears that dislocation damping plays a negligible role in causing absorption of ultrasonic waves in this case. Mason and Bateman[10.14] have suggested that a mechanism initially proposed by Akheiser may be responsible. The total thermal motion of the atoms in a solid may be represented as the combined effect of number of longitudinal and shear waves coupled to each other by non-linear terms in the force–distance relation operating between adjacent atoms. If a strain is suddenly applied to the solid a temperature change will ensue. It is assumed that each type of wave, longitudinal or shear, will experience a different temperature change depending upon the type of wave and its direction. These differing temperature changes will relax in a short time interval to a common temperature change. The total relaxation time for this process in the presence of an ultrasonic wave will be the time required to transfer the ultrasonic wave, regarded as a very low energy phonon, into thermal phonons plus the time required to equalize the temperature differences between phonons.

10.6 Phonon–Electron Interactions

If an ultrasonic wave travels through a piezoelectric medium the instantaneous sinusoidal strain distribution within the medium will be accompanied by a sinusoidal electrical field distribution. If the medium is also a semiconductor the current carriers will be influenced by the sinusoidal electrical field and can either extract or donate energy to the ultrasonic waves so that the piezoelectric semiconducting medium can act as either an ultrasonic attenuator or amplifier. The most frequently used material to date has been cadmium sulphide which is both piezo-electric and photoconductive. If the cadmium sulphide is unilluminated it displays little attenuation but if it is illuminated so that thereby current carriers are generated, its attenuation increases. If an electrical field, parallel to the direction of propagation of the ultrasonic waves, is now applied the attenuation changes. As the electrical field is increased the attenuation increases to a maximum beyond which it falls rapidly, passes through zero and becomes negative, that is the attenuation has changed into amplification. The cross-over point between attenuation and amplification occurs when the current carrier velocity equals the acoustic velocity. Physically the ultrasonic amplifier behaves in a similar manner to the travelling-wave tube. The sinusoidal piezoelectric field travelling along the system axis tends to collect the current carriers into bunches so that at any given time there will be a spatial distribution of charge density within the medium. This distribution will travel through the medium with the current carrier drift velocity. If this drift velocity is less than the ultrasonic velocity the charge bunches will tend to lag behind the bunching field and thereby experience an accelerating force. The increase in carrier energy when it is accelerated can only come from the ultrasonic wave which must therefore be attenuated. With a drift velocity greater than the ultrasonic velocity the bunched carriers move ahead of the travelling electrical field and experience a retarding force. The loss in kinetic energy reappears as ultrasonic energy, that is ampli-fication occurs.

White[10.15] has derived an expression for the ultrasonic absorption or amplification which agrees reasonably well with experimental data. Following his method we start with the two piezoelectric equations involving the piezoelectric constant e given in table 3.1, viz.:

$$T = cS - eE \qquad \ldots \quad \ldots \quad \textbf{(10.5)}$$

$$D = eS + \varepsilon E \qquad \ldots \quad \ldots \quad \textbf{(10.6)}$$

where T and S denote the stress and strain respectively and E and D the electrical field strength and displacement respectively. We have ignored the superscripts of table 3.1 in order to simplify the presentation. Since $S = \dfrac{\partial \xi}{\partial x}$, where ξ denotes the particle displacement, we have on combin-ing equations (1.15) and (10.5) that:

$$\frac{\partial T}{\partial x} = \rho\left(\frac{\partial^2 \xi}{\partial t^2}\right) = c\frac{\partial^2 \xi}{\partial x^2} - e\frac{\partial E}{\partial x} \qquad \ldots \quad \ldots \quad \textbf{(10.7)}$$

Provided we can derive an expression for E in terms of ξ we can then obtain an expression for the effective elastic modulus from which, if it turns out to be complex, we can derive an expression for the ultrasonic attenuation. First of all we derive an expression for the electrical displacement D in terms of E and the properties of the semiconductor. We then use equation (10.6) to eliminate D and so obtain an expression relating E to S. In order to obtain D in terms of the semiconductor properties we make use of the following equations:

Gauss' equation

$$\frac{\partial D}{\partial x} = Q \qquad \qquad \text{(10.8)}$$

where Q denotes the space charge and is given by:

$$Q = -qn_s$$

in which we assume the current carriers to be electrons each of charge $-q$ and where n_s denotes the number of electrons giving a total space charge equal to Q.

Continuity equation

$$\frac{\partial J}{\partial x} = -\frac{\partial Q}{\partial t} \qquad \qquad \text{(10.9)}$$

where J denotes the current density which is given by:

$$J = q\mu n_c E + qD_n\left(\frac{\partial n_c}{\partial x}\right) \qquad \text{(10.10)}$$

where μ denotes the electron mobility, n_c the number of electrons in the conduction band and D_n is the electron diffusion constant. The first term on the right-hand side of equation (10.10) represents the drift of the electrons due to the applied field E and the second term the drift caused by the diffusion resulting from electron concentration gradients. The number of electrons in the conduction band, n_c, is related to the number required to produce electrical neutrality in the absence of an ultrasonic wave, n_0, and n_s by:

$$n_c = n_0 + fn_s \qquad \qquad \text{(10.11)}$$

where f is a coefficient relating to the trapping of electrons by holes. With no trapping $f = 1$.

We now combine equations (10.8) to (10.11) in order to eliminate J, n_s and n_c and find:

$$-\frac{\partial^2 D}{\partial x \partial t} = \mu\frac{\partial}{\partial x}\left[\left(qn_0 - f\frac{\partial D}{\partial x}\right)E\right] - fD_n\left(\frac{\partial^3 D}{\partial x^3}\right) \qquad \text{(10.12)}$$

Assuming that we are dealing with small signals, we can write:

$$E = E_0 + E_1 e^{j(\omega t - kx)} \qquad \text{(10.13)}$$

and

$$\xi = \xi_0 e^{j(\omega t - kx)} \quad \ldots \ldots \quad (10.14)$$

where k denotes the complex propagation constant given by:

$$k = \beta - j\alpha \quad \ldots \ldots \ldots \quad (10.15)$$

where, as usual $\beta = \omega/c_0$, c_0 denoting the ultrasonic velocity, and α the absorption coefficient.

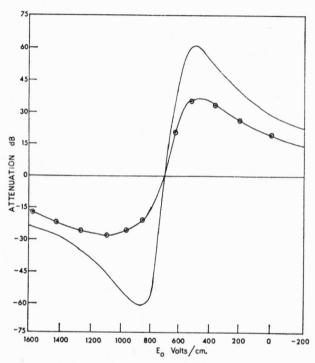

Fig. 10.3 Attenuation in illuminated cadmium sulphide at 45 mc/s as a function of the applied drift field. ——⊙——— experimental data, ———————— theoretical attenuation. (After White (10.15).)

If we now substitute equations (10.6), (10.13) and (10.14) into equation (10.12) we obtain:

$$E_1 = \frac{jk\xi_0 e}{\varepsilon}\left\{1 - \frac{j\sigma}{\varepsilon\omega}\left[1 + \mu f E_0 \frac{k}{\omega} - j\frac{k^2}{\omega}D_n f\right]^{-1}\right\}^{-1} \quad (10.16)$$

where $\sigma = \mu q n_0$ and is the semiconductor's conductivity.

We now substitute equations (10.14) and (10.16) into equation (10.7) to obtain an expression for the effective elastic constant c', viz.:

$$c' = c\left\{1 + \frac{e^2}{\varepsilon c}\left[1 - \frac{j\sigma}{\varepsilon\omega}\left(1 + f\mu\frac{k}{\omega}E_0 - jD_n f\omega\left(\frac{k}{\omega}\right)^2\right)^{-1}\right]^{-1}\right\} \quad (10.17)$$

We can now derive an expression for the absorption coefficient from c' since (cf. equation 6.4):

$$\frac{c'}{\rho} = \frac{\omega^2}{(\beta - j\alpha)^2} \qquad \ldots \ldots \quad \textbf{(10.18)}$$

Under the assumption that $|\alpha| \ll \dfrac{\omega}{c_0}$ equations (10.17) and (10.18) lead to:

$$\alpha = \frac{e^2}{2\varepsilon^2 c} \frac{\sigma}{c_0 \gamma} \left[1 + \frac{\sigma^2}{\gamma^2 \varepsilon^2 \omega^2} \left(1 + \frac{\varepsilon \omega^2 f^2 D_n^2}{\sigma c_0^2} \right)^2 \right]^{-1}. \quad \textbf{(10.19)}$$

where $\gamma = 1 - \left(\dfrac{f\mu E_0}{c_0} \right)$. The absorption coefficient is positive or negative depending upon whether $f\mu E_0$ is less than, or greater than c_0. If $f = 1$, $f\mu E_0$ is simply the electron drift velocity.

The agreement between theoretical prediction and experimental results is quite reasonable in view of possible inhomogeneities in the sample photoconductivity, in its measured piezoelectric constant and in the experimental difficulties involved in measuring the steep parts of the gain versus drift field curve. One set of results is shown in Fig. 10.3.

10.7 Phonon–Spin Wave Interactions

The magnetic properties of a ferromagnetic medium are due to the spin properties of the electrons of the individual atoms in the crystal lattice. These spin waves interact with the lattice spacings and can give rise to shear waves with frequencies of the order of several thousands of megacycles per second. Normally the attenuation of the shear waves at these frequencies is extremely high at room temperatures so that originally it was thought that any work in this field would have to take place at very low temperatures. However, the absorption of certain rare earth garnets, which are ferromagnetic, is very low at room temperatures[10.16] and this discovery has made possible the development of devices utilising the interaction between spin and shear waves to produce known signal delays at microwave frequencies. Such devices find their greatest application in the processing of radar information and there is considerable current interest in this field. The delay arises from the fact that the shear wave velocity is considerably less than that of the electromagnetic waves so that any device which converts electromagnetic energy into shear wave energy and then reconverts it into electromagnetic energy will introduce a large signal delay.

The arrangement often used is shown in Fig. 10.4 in which a short wire electromagnetic radiator is fed from a suitable waveguide or co-axial system and placed close to one end of an axially magnetised rod of yttrium iron garnet (y.i.g.). The transverse magnetic field from the radiator excites magnetostatic modes in the y.i.g. rod. These modes are termed magnetostatic since because the rod dimensions are much smaller than the electromagnetic wavelength, the resulting time varying magnetic field within the rod can be calculated from the rod geometry neglecting wave propagation effects. Under suitable conditions these magnetostatic modes are converted into spin waves which, in turn and again under suitable conditions are converted into shear acoustic waves. The conversion from one energy form into another requires that the axial magnetic field within the y.i.g. rod varies along the rod length. This is easily achieved if the rod is magnetised by an external magnetic field when because of the depolarising effects of the rod ends the internal magnetic field is approximately parabolic with a maximum at the rod centre.

The theory which we shall now present can only be incomplete since far too much space would be required to give a thorough treatment. Giving an incomplete treatment, however, poses two problems. In the first place certain statements must be made which the reader is asked to

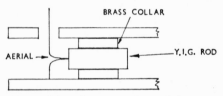

Fig. 10.4 Apparatus for studying the conversion of electromagnetic into acoustic energy by means of spin waves. (After Strauss (10.23).)

take on trust and in the second place all the available literature uses the C.G.S. system of units. There seems to be little point in confusing the reader who wishes to go on to the literature by using the M.K.S. system merely for the sake of uniformity throughout this book. Thus in what follows we shall be using the C.G.S. system.

The dispersion relationship, giving the relation between wavelength and frequency, for propagation parallel to the axial magnetising field for both the magnetostatic and spin waves is[10.17]:

$$\frac{\omega}{\gamma} = H_i + \frac{A}{k^2} + Dk^2 \quad . \quad . \quad . \quad (10.20)$$

where ω denotes the angular frequency, γ the gyromagnetic ratio (the ratio of the magnetic to the angular moment for an electron), H_i the internal magnetic field, k the wavenumber (given by $2\pi/\lambda$), A is a constant for the rod defining a particular magnetostatic mode and D the exchange energy constant. For the lowest order mode, A is given[10.17] by $2\pi M\left(\dfrac{2\cdot405}{r}\right)^2$ for a rod of radius r and[10.18] by $2\pi M\left(\dfrac{\pi}{b}\right)^2$ for a rod of rectangular cross-section $a \times b$ where a is very much greater than b. M denotes the saturation magnetisation. The term A/k^2 arises from the interaction of any one dipole with all other dipoles and the term Dk^2

arises from the quantum mechanical exchange energy between two adjacent dipoles[10.17]. If the exchange term is negligible, as when k^2 is small, the internal time varying magnetic field is represented by the magnetostatic modes but when the exchange term predominates, at large values of k^2, the time varying magnetic field is represented by spin waves. Spin waves can be pictured as sinusoidal time and spatial variations in the spins of the atomic magnetic dipoles about the axial magnetising field. In general the time varying magnetic field is a combination of magnetostatic and spin waves.

For y.i.g. we have $D = 5 \ 10^{-9}$ Oe. cm^2 and $2\pi M = 875$ Oe. Thus for a rod 3 mm in diameter energised at a frequency of 10^9 c/s, equation (10.20) becomes:

$$10^2\pi = H_i + \frac{2 \cdot 24 \ 10^5}{k^2} + 5 \ 10^{-9}k^2 \quad . \quad . \quad \textbf{(10.21)}$$

If we assume, for example, that at the rod end face adjacent to the radiator the internal magnetic field is 100 Oe., equation (10.21) gives $k = 2 \ 10^5$ or 33·5 with corresponding wavelengths of $3 \cdot 14 \ 10^{-5}$ and 0·19 cm respectively. The electromagnetic wavelength is closer to the longer wavelength and therefore the wave corresponding to $k = 33 \cdot 5$ will be preferentially excited. Thus close to the end face the magnetostatic wave predominates. However, as we move away from the end face towards to centre of the rod, H_i increases and therefore k increases so that the exchange term Dk^2 starts to become comparable with A/k^2, that is the energy of the magnetostatic mode transfers into spin-wave energy, particularly when the two roots of equation (10.20) coincide, that is when $k_{\text{magnetostatic}} = k_{\text{spin}}$. For our numerical example this occurs when:

$$k^2 = \sqrt{A/D} = 6 \cdot 68 \ 10^6 \quad . \quad . \quad . \quad \textbf{(10.22)}$$

i.e. $k = 2 \cdot 58 \ 10^3$

This position in the rod has another significance in that here the energy travelling from the radiator is reflected. The group velocity for the combined magnetostatic and spin wave travelling from the radiator is given by:

$$\frac{d\omega}{dk} = -\frac{2A}{k^2} + 2Dk = 2\left(\frac{D}{A}\right)^{\frac{1}{4}}[\sqrt{AD} - \sqrt{AD}] = 0 \quad \textbf{(10.23)}$$

Since energy travels at the group velocity it must be reflected when the group velocity falls to zero otherwise it would just pile up. This point is known as the turning point. The internal magnetic field at the turning point, denoted by $H_i(y)$ can be found by combining equations (10.20) and (10.22) to be given by:

$$\frac{\omega}{\gamma} - H_i = 2\sqrt{AD} \quad . \quad . \quad . \quad . \quad \textbf{(10.24)}$$

For our numerical example $2\sqrt{AD} = 67$ mOe which is much less than $H_i(y)$ which equals $\pi 10^2$ Oe. For this reason the condition for the turning point is often quoted as:

$$H_i(y) = \frac{\omega}{\gamma} . \qquad . \quad . \quad . \quad . \quad . \quad (10.25)$$

As the reflected spin wave travels back towards the end face it is moving in a region of decreasing internal magnetic field and therefore its wavenumber k must increase. Spin waves depend upon the exchange forces between adjacent atoms in the crystal lattice. Since these forces are essentially electrostatic they will involve the interatomic distance and any variation in one will affect the other. Thus a spin wave will give rise to elastic strains within the medium and when the spin wave-number equals the elastic wave wavenumber an elastic wave will be launched[10.19], in fact it will be a shear wave propagating in the same direction as the spin wave. The point within the rod at which the shear wave develops is known as the cross-over point. Denoting this point by y_1 we have:

$$k^2 = \left(\frac{\omega}{\gamma} - H(y_1)\right)\Big/ D = \frac{\omega^2}{c^2} \quad . \quad . \quad . \quad (10.26)$$

where c denotes the shear wave velocity. For y.i.g. $c = 3 \cdot 76 \; 10^5$ cm sec^{-1}, hence $k = 1 \cdot 7 \; 10^4$ and $\frac{\omega}{\gamma} - H(y_1) = 1 \cdot 45$ Oe.

The theory outlined in the preceding paragraph is only approximately correct since the spin and elastic waves cannot properly be separated and should really be considered as a coupled system of spin and elastic waves known as a magnetoelastic wave. This point has been considered by Lacklison and Lewis[10.20]. Taking the dispersion equation for a magnetoelastic wave given by Schlomann and Joseph[10.21]:

$$\left[k^2 - \left(\frac{\omega}{c}\right)^2\right]\left[k^2 - \frac{1}{D}\left(\frac{\omega}{\gamma} - H_i\right)\right] = \frac{b^2}{DMc_{44}} . \quad (10.27)$$

where c_{44} is an elastic constant and b a magnetoelastic coupling constant, Lacklison and Lewis calculated numerically the time taken for the wave to travel from the turning point to cross-over. They found that this time is some 34% less than that given by the approximate theory assuming pure spin propagation but that the time from the turning point to well beyond cross-over agreed well with that from the approximate theory. Thus two errors in the approximate theory tend to cancel out.

The efficiency of coupling between electromagnetic and spin energy and between spin and elastic energy is of importance in the design of microwave delay units. According to Schlomann and Joseph[10.22] the former depends inversely upon the square root of the magnetic field gradient at the turning point while the latter depends directly upon the field gradient at the cross-over point.

References

10.1 Mason, W. P., *Physical Acoustics and the Properties of Solids*, p. 202. Van Nostrand (1958).

10.2 Stern, R., Natale, G. G. and Rudnik, I., *Proc. 5th Int. Cong. Acoust.*, C.31 (1965).

10.3 Mason, W. P., *Physical Acoustics and the Properties of Solids*, p. 200. Van Nostrand (1958).

10.4 Zener, C., *Phys. Rev.*, **53**, 90 (1938).

10.5 Mason, W. P., *Physical Acoustics and the Properties of Solids*, p. 208. Van Nostrand (1958).

10.6 Mason, W. P. and McSkimin, H. J., *J. Appl. Phys.*, **19**, 940 (1948).

10.7 Huntingdon, H. B., *J. Acoust. Soc. Amer.*, **22**, 362 (1950).

10.8 Mason, W. P., *Physical Acoustics and the Properties of Solids*. Van Nostrand (1958).

10.9 Bozorth, R. M., *Ferromagnetism*. Van Nostrand (1951).

10.10 Snoek, J. L., *Physica*, **8**, 711 (1941).

10.11 Ké, T. S., *Phys. Rev.*, **74**, 9 (1948).

10.12 Zener, C., *Elasticity and Anelasticity of Metals*, p. 111. University of Chicago Press (1948).

10.13 Mason, W. P., *J. Acoust. Soc. Amer.*, **32**, 458 (1960).

10.14 Mason, W. P. and Bateman, T. B., *J. Acoust. Soc. Amer.*, **36**, 644 (1964).

10.15 White, D. L., *J. Appl. Phys.*, **33**, 2547 (1962).

10.16 Olson, F. A., Yaeger, J. R. and Wilson, R. A., *Proc. 5th Int. Cong. Acoust.*, D7 (1965).

10.17 Fletcher, P. C. and Kittel, C., *Phys. Rev.*, **120**, 2004 (1960).

10.18 Schlomann, E., *Proc. I.E.E.E.*, **53**, 1495 (1965).

10.19 Kittel, C., *Phys. Rev.*, **110**, 836 (1958).

10.20 Lacklison, D. F. and Lewis, M. F., *Electronics Letters*, **2**, 378 (1966).

10.21 Schlomann, E. and Joseph, R. I., *J. Appl. Phys.*, **35**, 2382 (1964).

10.22 Schlomann, E. and Joseph, R. I., *J. Appl. Phys.*, **35**, 167 (1964).

10.23 Strauss, W., *J. Appl. Phys.*, **36**, 118 (1965).

11

Miscellaneous Applications
of Ultrasonics

11.1 Introduction

At one time or another ultrasonic vibrations have been applied to nearly every known physical and chemical process and a number of claims as to their effectiveness have been made. With the passage of time, however, the useful applications have become better defined while many others have been dropped. In this chapter we shall outline a few of the more useful low power applications; high power applications have been considered in chapter 5.

11.2 Flaw Detection

At any physical discontinuity in a medium there is likely to be an acoustical discontinuity which will reflect ultrasonic waves. If, therefore, a pulse of ultrasonic energy is transmitted into the medium from a transducer, the time taken for this pulse to travel from the transducer to the discontinuity and back again will give a measure of the distance of the discontinuity provided that the acoustic velocity of the medium is known. In addition the magnitude of the echo will give an indication of the size of the discontinuity. The types of discontinuity investigated with this technique range from flaws in solids to tumours within human bodies and shoals of fish in the sea. In its simplest form a flaw detector consists of an arrangement similar to that shown in Fig. 11.1. A pulsed oscillator, controlled by a pulse generator, applies a pulse modulated radio frequency voltage to a transducer which emits pulse modulated ultrasonic waves. These travel out from the transducer until they meet a discontinuity where part of the ultrasonic energy is scattered. Some of this scattered energy arrives back at either the same transducer or a separate receiver where it is converted back into an electrical signal, amplified and then displayed on some suitable display unit such as a pen recorder, or, more commonly, an oscilloscope. We shall now consider the various factors which have to be taken into account in the design of a flaw detector starting with the electronic section.

The pulse repetition frequency is governed by the time taken by the ultrasonic waves in travelling out to, and back from, the farthest flaw it is proposed to detect. The time interval between the pulses must exceed this travel time otherwise a received echo pulse may appear on the display unit after the next pulse has been emitted and appear to indicate an erroneously near flaw. In practice the distance of the remotest detectable flaw is fixed either by the dimensions of the body under investigation or by its absorption of the ultrasonic energy in relation to the magnitude of the transmitted energy and the receiver sensitivity. The maximum distance at which a given flaw can be detected will increase with an increase in the transmitted power, an increase in the receiver sensitivity and a decrease in the medium's absorption.

The duration of the pulse is fixed by the minimum separation between two flaws along the direction of propagation of the ultrasonic energy which are to be resolved. If the pulse duration exceeds twice the time

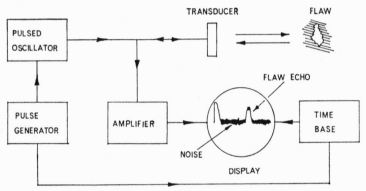

Fig. 11.1 Simplified block diagram of a flaw detector.

taken for the pulse to travel between the two flaws the two echoes will overlap and so cannot be completely resolved.

While it may be advantageous to have as sensitive a receiver as possible the actual sensitivity will be limited by the presence of electrical noise generated within the receiver itself, predominantly in the first stage of amplification, since this noise signal will be amplified along with any echo signal by the subsequent stages whose noise contribution will be effectively reduced since it will undergo less amplification than that from the first stage. Unless special integrating techniques are used, echo signals of an amplitude equal to or less than the noise signal cannot be detected. Since the magnitude of the noise signal at the amplifier output is proportional to the square root of the amplifier bandwidth, the noise can be reduced by reducing the bandwidth. However, if the bandwidth is reduced excessively the rise time of the echo signal at the amplifier output becomes too long so that the signal cannot reach its maximum before the end of the pulse arrives and, therefore, the signal becomes attenuated thereby reducing the signal to noise ratio. In addition the slow rise of the leading edge of the pulse reduces the accuracy with which the flaw can be located—this may or may not be important depending upon whether the equipment is being used to locate a flaw accurately or to just indicate its presence. If the pulse lasts for T microseconds it is common practice to make the amplifier bandwidth about $2/T$ mc/s. If only one ultrasonic frequency is to be used the amplifier can be a simple straight amplifier although since in many applications differing frequencies are used, a superheterodyne receiver is more common. Even if two transducers are used, one to transmit and one to receive, some of the electrical energy applied to the transmitter will inevitably 'leak' to the receiver and will probably cause it to overload. When this occurs the amplifier becomes non-linear and its various capacitances, such as the coupling and decoupling capacitors, will charge up. These will take some time to discharge at the end of the transmitted pulse and during this time the receiver will be paralysed and unable to amplify any echo signals; thus flaws close to the transmitter will not be detected. The time for which the receiver is paralysed can be reduced by making the time constants associated with these capacitances as small as possible. Because of the absorption in the medium, signals from distant flaws will be weaker than those from nearer flaws and therefore the amplifier gain is sometimes swept so that it increases with time as measured from the instant at which the transmitted pulse is emitted. Thus two equal flaws, situated at differing distances, will give approximately equal signals at the amplifier output.

The function of the display unit is to present in visual form the echo signals in such a way as to indicate the position and magnitude of the flaw. In straightforward flaw detection systems the commonest display takes the form of a cathode ray tube A scan which is identical to that used with pulse techniques applied to the measurement of the absorption of liquids described in section 8.8. The distance of the flaw echo along the trace is proportional to the distance of the flaw from the transmitter and the height of the echo signal is a measure of the size of the discontinuity presented to the incident ultrasonic energy by the flaw. This discontinuity is only partially related to the physical size of the flaw since it also depends upon the acoustic mismatch, shape and orientation of the flaw. Alternative types of display will be considered when we deal with medical applications of flaw detection.

The transducer mounting, or probe, which is used depends upon the orientation and position of the flaw relative to some accessible surface against which the probe is placed and also upon the shape of the body being investigated. If we assume that the accessible surface is flat, the main reflecting surface of the flaw can be either parallel or at an angle to the flat surface. In the first case, and provided that the flaw is not so close to the surface that its echo arrives during the time in which the amplifier is paralysed, a simple probe of the type illustrated in Fig. 11.2a can be used as both the transmitting and receiving probe if it is placed directly above the flaw. In the second case an angled probe (Fig. 11.2b) is used in order that the flaw presents the maximum reflecting surface to the incident ultrasonic energy. If the body being investigated is a plate an alternative technique is used in which the longitudinal waves from the transmitter are caused to meet the plate front surface at a

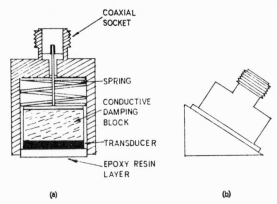

Fig. 11.2 Typical probes for use in flaw detection.

suitable angle such that shear waves, formed by mode conversion, are refracted into the plate (cf. section 1.14). These shear waves travel within the plate by a series of reflections, as shown in Fig. 11.3 until they meet and are reflected by the flaw either back to the transmitting probe or, alternatively, to a suitably positioned receiving probe. Since the velocity of shear waves is less than that of longitudinal waves, their use improves the resolution of the system. If the transmitting probe is angled such that the refracted shear waves travel parallel to the plate's front surface, surface defects can be detected; such waves can also be used to

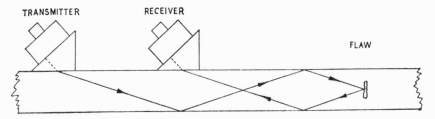

Fig. 11.3 Flaw detection in a plate using angled probes.

detect flaws parallel to the front surface but which are too close to it to be detected by the probe shown in Fig. 11.2a. The region between such a flaw and the near front surface of the body can be regarded as a plate. According to Lamb[11.1] a plate will support travelling waves in which the energy is propagated by the plate bending sinusoidally with both time and distance. Since the velocity of these Lamb waves is proportional, *inter alia*, to the square root of the plate thickness, the wave velocity will differ from that of the incident waves and thus the resulting acoustic discontinuity will reflect some of the incident energy to give an echo signal. The condition that Lamb waves be produced is, for example, for Zirconium[11.2], that the product of the ultrasonic frequency and the plate thickness be less than $2 \cdot 10^6$ cm c/s.

There are one or two practical points associated with the design of probes which are worth mentioning. The transducer should be loaded at its rear surface by a medium of large specific acoustic impedance in order to reduce its mechanical Q sufficiently to give a good pulse response. This rear loading medium should also have a high absorption so that ultrasonic waves reflected from its rear surface do not give rise to false echoes. An epoxy resin, which bonds well to the transducer surface, containing dispersed tungsten powder fulfils both requirements adequately. The probe should naturally be robust and if it is of the type shown in Fig. 11.2a the transducer front surface should be protected against abrasion by a coating of either plastic or epoxy resin. Whichever type of probe is used it must be acoustically coupled to the body under test by means of an oil film.

Since the far side of a plate may be regarded as a large flaw, flaw detection methods can be used to measure the plate thickness. This technique has the great advantage that access to one side only of the plate is required.

It is implicit in the foregoing discussion of flaw detection that since the position of the flaw may only be known approximately, if at all, considerable skill is required by an operator if all important flaws are to be discovered and if artifacts due to beam spreading and stray reflections from the sides of the body are to be recognised as such.

If the accessible surface of the body being investigated is not flat, it is not possible to place the probe directly on to the surface and in this case an immersion technique has to be used in which the body and the probe are immersed in a fluid, usually water, with the probe suitably angled but at a distance from the body surface. As an illustration of this technique we consider the equipment used to examine small bore metal tubes such as are used for canning nuclear reactor fuel. These tubes must be examined for small defects which, if undetected and the tube accepted, will limit the useful life of the fuel element. One such examination system is shown in outline in Fig. 11.4. The tube is simultaneously rotated and moved axially through a water bath whose sides contain two bushes through which the tube passes. Two ultrasonic probes are used, each acting as both transmitter and receiver, one of which detects longitudinal flaws and the other circumferential flaws. The main lobe from the first probe is angled such that its axis does not pass through the tube's axis so that when its radiation meets the tube outer surface it is partially reflected and partially refracted into the tube wall and sets up either shear or Lamb waves which travel around the tube until reflected back to the probe from a flaw. The exact mechanism of wave propagation in this system is not yet fully understood. The second probe, detecting circumferential flaws, has its main lobe axis passing through the tube axis but the beam makes an angle with the tube surface so that energy coupled into the tube wall can travel parallel to the tube axis until reflected back to the probe from a flaw. The output from each probe consists of a train of pulses, at the pulse repetition frequency, occurring whenever a flaw passes through the ultrasonic beam within the tube. These echoes can be displayed on an A scan but here a pen recorder is more convenient. Using 6 mc/s focused probes in a similar

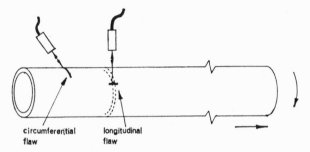

circumferential longitudinal
flaw flaw

Fig. 11.4 Elements of a continuous tube inspection system.

system, Rooney and Reid[11.3] found that for best results the angle of incidence of the radiation on to the tube wall is fairly critical and is a function of both the tube diameter and the wall thickness. In addition they found that a defect on the outer wall gave rise to two echoes of differing amplitudes whereas a defect on the internal wall gave only one echo. For identical defects, the second outer wall echo is approximately of the same amplitude as that from the inner wall defect and they therefore arranged an electronic gate to suppress the first outer wall echo since, as well as locating defects, they were also interested in comparing defect sizes.

11.3 Medical Applications of Flaw Detection

For many years the main technique available to the clinical diagnostician for the internal examination of the human body not requiring an operation under anaesthesia has been that provided by X-rays. However, the use of X-rays has two disadvantages in that, first, they do not delineate satisfactorily the soft tissues of the human body and, secondly, they are not advisable during the early months of pregnancy. Since differing tissues possess differing acoustic impedances and absorptions Dussik[11.4] attempted in 1942 to apply ultrasonic techniques to the diagnosis of abnormalities within the human body. Unfortunately he tried to use a transmission technique in which he recorded the variation in the intensity of an ultrasonic beam after it had completely traversed the human head presenting the information very much as an X-ray photograph. This technique proved abortive because of the high absorption and reflection by the skull. Success with ultrasonics as a diagnostic tool only came with the application, initially by Wild[11.5], of pulse echo flaw detection systems. These rely on the differing acoustic impedances of differing tissues since even a small impedance difference will give a tolerable echo. The degree of sophistication required in an instrument for medical applications will depend upon the information required by the diagnostician. For example, if the patient is suspected of having suffered brain damage from a blow an indication of a shift in the mid-line of the brain may well be sufficient, whereas in gynaecological applications a detailed cross-section, or tomogram, is of more use. In the first case a simple A scan display suffices but the second case requires a more complicated compound scan. Since a simple flaw detector type of instrument suffices for the first case we shall restrict the present discussion to compound scan equipments.

Ideally an ultrasonic tomogram presents a picture of what would be seen if the patient were to be sliced completely across a plane lying, for example, at right angles to the spine. In principle this could be achieved by placing a probe against the patient at a suitably chosen position and varying the angle at which the radiation entered the body. The trace on a cathode ray tube could easily be made to follow the angle made by the probe beam relative to some fixed axis and the echoes displayed by only allowing the cathode ray tube electron beam to strike the phosphor screen when an echo is received. This technique, however, will not give satisfactory results since a discontinuity within the body may happen to present only a small reflecting surface to the radiation coming from the probe. If, however, the probe were to be situated at a different position, a better echo would be obtained. It is necessary, therefore, that the probe moves around the body in a chosen plane. This condition immediately gives rise to the problem of how to couple the radiation from the moving probe into the body. Two techniques have been developed for this which, for convenience, we may label as the wet and the dry techniques: the wet technique couples the probe to the patient by means of a water bath in which, or against which the patient is placed whereas in the dry technique the probe is in direct contact with the patient's skin via a thin coupling film of olive oil. Both techniques give comparable results, although the wet technique gives slightly better results but is less acceptable by the patient.

As an example of the wet technique we consider Howry's equipment. In his first equipment Howry[11.6] sat the patient in a water-filled circular tank. The probe, situated at a suitable height, was suspended from a carriage mounted on a straight track which formed a chord of the circular top of the tank. This track moved around the top of the tank and at the same time the probe carriage moved backwards and forwards to give a scan illustrated in Fig. 11.5a so that over the central portion of the tank any reflecting surface was scanned from several angles. The movement of the trace across the face of the cathode ray tube used to display the echoes was made to follow the beam of radiation from the probe by a combination of mechanical and electronic linkages and the resulting picture recorded on film. Howry found that in the majority of cases the full 360° scan was unnecessary and could be replaced with little loss by one of 180° or less. In his second apparatus, therefore, he[11.7] used a semi-circular tank with the metal section at the centre of the straight side replaced by a thin plastic sheet against which the patient was pressed. This thin plastic sheet was effectively transparent to the ultrasonic waves and provided that the patient was coupled to the sheet by a thin film of oil and all air bubbles removed good results were achieved.

Using a similar technique with the patient's face pressed against an opening in the side of a water-filled tank, Baum and Greenwood[11.8] obtained excellent tomograms of the human eye.

The main work using dry techniques has been carried out by Donald and Brown[11.9] who developed a sophisticated scanning system primarily intended for gynaecological investigations. A focused transducer was

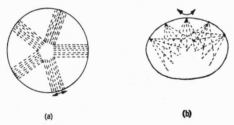

(a)   (b)

Fig. 11.5 Illustration of two types of scan for obtaining tomograms. (*a*) Howry's first technique; (*b*) Donald and Brown's technique.

mounted in a housing suspended by a system of rods from an overhead arm. These rods controlled the pressure and angle at which the probe pressed against the abdomen of a prone patient. The probe was placed in position and rocked over an angle of about 90° to scan a sector of the patient. The probe then moved automatically to a new position and the scan repeated. In this way a complete tomogram could be built up from a sequence of sector scans. The scanning system is shown in Fig. 11.5b.

In order that a meaningful picture of the interior of a patient be obtained it is necessary that the system possess adequate resolution. We have already considered the question of longitudinal resolution when dealing with flaw detection but ultrasonic tomography requires, in addition, good angular resolution which implies that the ultrasonic beam be as narrow as possible. In principle the beam width can be made small by using a small diameter transducer at a sufficiently high frequency that the beam divergence is small. Unfortunately since the absorption by human tissues is of the order of 2 dB/cm/mc/s and the maximum permissible ultrasonic power into a human being is limited by heating and cavitation effects, the use of high frequencies is only possible if the required depth of penetration is restricted to a few centimetres. In order to penetrate completely through the human body the frequency cannot exceed a few megacycles per second which implies that the transducer diameter must be of the order of one centimetre. The beam width will be the same or greater than the transducer diameter except at the focus of a focused transducer. Hence as the beam sweeps across a target, for example a thin wire, the reflections from the wire would indicate that the wire is one beam width wide, thus the system's angular resolution will be poor. There are two ways of improving angular resolution. When the target is scanned the peak echo will appear when the target is on the axis of the main lobe of the radiation. If, as is usual, the target is scanned from various positions and the tomogram is built up by intensity modulation a maximum in light output from the face of the cathode ray tube will occur over a region corresponding, approximately, to the shape and size of the target provided there is perfect correlation between the cathode ray tube trace and the angular position of the transducer. This light maximum can be emphasised by a suitable choice of amplifier gain and the aperture of the camera used to photograph the tomogram although at the expense of losing some detail from weakly reflecting surfaces. An alternative technique, still in the early stages of development, uses a digital computer to analyse the tomogram and to correct for the finite beam width.

All techniques using compound scan require that the positioning and direction of the trace on the cathode ray tube face correspond exactly with the position and angle of the probe. While this can be achieved with a combination of mechanical and electrical devices a straightforward mechanical system does, at the moment, lead to better results mainly because of the drift in electronic component values with both time and temperature. A degradation in performance shows itself as a reduction in resolution.

11.4 Sonar

Sonar is the name given to the use of radar techniques in which ultrasonic waves replace electromagnetic waves and both flaw detection and the medical applications discussed in the preceding section are special cases of general sonar techniques. Wartime applications of sonar are mainly concerned with submarine location but peaceful uses are those such as depth finding and fish location. The principles are almost identical to those already described in this chapter. For straightforward depth finding a simple form of 'sing around' system, somewhat similar to Greenspan's method for measuring ultrasonic velocities (cf. section 8.8) can be used in which the pulse repetition frequency of the system becomes inversely proportional to the depth. The difficulty in using sonar for locating fish shoals is that since many desirable fish swim close to the sea bed the region of interest on a conventional display is cramped near to one end. Hopkin and Haslett[11.10] have described one ingenious system for overcoming this disadvantage in which the display is locked to the sea bed echo, that is it does not start until the sea bed echo is received. Since the shoal echoes arrive before that from the sea bed, they are stored on a continuously rotating magnetic drum and released only after the arrival of the sea bed echo. By this means they arrange to display only echoes from the region lying within ten fathoms of the sea bed irrespective of the actual depth of the sea.

11.5 Blind Guidance

Although ultrasonic waves have a restricted useful range in air many attempts have been made to use sonar techniques to provide a guidance system for the blind. The main difficulties arise from the necessity to make the equipment sufficiently small and light to be portable and in presenting the information to the blind person in a meaningful manner. Recently Kay[11.11] has described a system which appears to work well and which has been well received by the blind. He uses two dielectric microphones, one to transmit and the other as a receiver. The transmitter radiates acoustic waves with a frequency which decreases linearly with time from 60 kc/s to 30 kc/s, rises rapidly back to 60 kc/s and then falls again and so on. Signals reflected back from an obstacle are picked up by the second microphone and mixed with the outgoing signal to produce a difference frequency which is passed through a low pass filter, rejecting any signal with a frequency greater than 3 kc/s before being amplified and fed into a pair of head-phones. The frequency of the signal in the head-phones will depend upon the rate of change of frequency of the transmitted ultrasonic waves and the distance of the obstacle. Thus with reasonable directivity of the microphones a blind person can detect obstacles and estimate their distance away. The apparatus described by Kay uses a rate of change of frequency of 50 kc/s/sec and has a maximum range of 20 ft. In use it has been found more useful for identifying landmarks rather than as a simple guidance system.

11.6 Delay Lines

Ultrasonic delay lines have found applications in early digital computers as storage registers but are now used in colour television receivers and in processing radar information. As their name implies they are used to delay signals. Electrical signals propagate with a finite speed along wires and through electrical networks and therefore these can be used to provide a delay. However, the delays obtainable from purely electrical delay lines are small, ranging from 40 nanoseconds per metre length for some co-axial cables up to 10 microseconds per section for a lumped circuit delay line. Often these delays are insufficient or the frequency response of the delay circuit is poor and to achieve longer delays with good pulse response advantage is taken of the relatively slow moving ultrasonic wave. A simple delay line consists of two transducers connected at opposite ends of a low loss rod. Electrical signals fed into one transducer at its resonant frequency are converted into ultrasonic waves, travel along the rod to the other transducer where they reappear as electrical signals. Using longitudinal waves and a fused quartz rod the typical delay is of the order of two microseconds for each centimetre of path length. If mode conversion techniques are used so that shear waves are propagated the delay is approximately doubled. If very long delays are required the quartz, or glass, rod is effectively folded back on itself to form a polygon[11.12] so as to give an effective long path length in a reasonably small volume as shown in Fig. 11.6. While this technique is

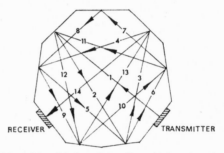

Fig. 11.6 Solid multi reflection delay line. (After Arenberg (11.12).)

capable of providing a wide bandwidth delay line with delays of up to 5 milliseconds it calls for considerable care in the manufacture of the polygon if its losses are not to be excessive and the delay stable with temperature. For even longer delays but with a restricted bandwidth, wire delay lines are used. Two magnetostrictive transducers are attached to each end of a long steel wire so that when a pair is energised by an electrical signal torsional waves are generated which travel along the wire to the other end where they are converted back into an electrical signal. With this device delays of up to 20 milliseconds have been achieved.

11.7 Ultrasonic Image Convertors

The ultrasonic image convertor is a device for producing a visible image of the ultrasonic pressure distribution across a wavefront. If an initially uniform wavefront is transmitted through a body containing discontinuities these will scatter some of the incident energy and therefore the wavefront emerging from the body will contain information about the interior of the body. If the discontinuities are sufficiently marked and the wavefront falls on to the image convertor, the visible image derived from the convertor will give a picture of the ultrasonic properties of the body hence the convertor will find its main application as a flaw detector.

Pohlman[11.13] developed one successful image convertor by using the

ultrasonic wavefront to orientate minute non-spherical particles suspended in a fluid medium where the degree of orientation was dependent
upon the ultrasonic pressure amplitude at any point. The differing degrees of orientation of the particles were made visible by illuminating
the particles. Recently, however, the piezoelectric image convertor has
become feasible. This type was first proposed by Sokolov and has been
brought to a viable state by Smyth, Poynton and Sayers[11.14] and
marketed commercially by Twentieth Century Electronics as camera
tube type UC9. This type of convertor is illustrated in Fig. 11.7. It

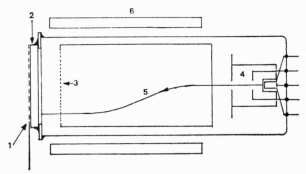

Fig. 11.7 Ultrasonic image convertor tube.

1. Signal plate.	4. Electron gun.
2. Quartz disc.	5. Electron beam.
3. Ion trap.	6. Focusing and deflection coils.

(After Smyth *et al.* (11.14).)

consists of a magnetically focused and deflected cathode ray tube with
the glass face replaced by an unplated disc of X-cut quartz. Immediately
behind the quartz disc is a fine mesh, connected to the final anode of the
electron gun, which serves as an ion trap while in front of the disc is
placed another mesh known as the signal plate. The ultrasonic radiation
falls on to the quartz disc from the signal plate side. If no ultrasonic
energy falls on to the quartz disc and it is scanned across its rear face
by the electron beam, as in a conventional television camera tube, the
charge distribution over the disc surface builds up negatively from electrons deposited by the electron beam until the surface potential equals
that of the electron gun cathode when no further electrons will be able
to reach the disc. If the ultrasonic beam is now switched on, the pressure variations across the face of the disc will set up corresponding voltage variations at the ultrasonic frequency. The electron beam now scans
the disc again and this time charge will be deposited at any point on the
disc surface during the positive going half cycle of the voltage variation
due to the ultrasonic waves. Sufficient charge will build up at each point
until the peak positive potential at that point does not exceed the
cathode potential. Thus a charge distribution will appear on the disc
surface corresponding to the ultrasonic pressure distribution. As the

electron beam scans across the disc surface and depositing charges on it, corresponding charges will be induced in the signal plate, that is, a current can be extracted from this plate which is proportional to the instantaneous rate of charge deposition. This current is amplified by a current amplifier whose output signal modulates the intensity of the electron beam in a normal cathode ray tube whose beam is deflected in synchronism with that in the image convertor. Thus a pictorial representation of the ultrasonic pressure distribution across the face of the quartz disc appears on the face of the cathode ray tube. Since the image convertor requires that the inner face of the quartz disc be at a uniform potential, equal to that of the cathode before a picture signal can be extracted, the ultrasonic beam is switched off after one frame has been scanned, the inner face potential allowed to go slightly positive by allowing positive ions to diffuse on to the disc surface and neutralise the excess negative charges, rescanning with the electron beam to restore the disc surface to cathode potential and the ultrasonic beam switched on again to produce a second picture frame. This sequence of events is repeated continuously to give a steady picture on the monitor cathode ray tube. Using a 2·5 cm diameter quartz disc at its resonant frequency of 4 mc/s, Smyth, Poynton and Sayers were able to obtain a resolution of 0·5 mm with a background noise level equivalent to an ultrasonic intensity of 10^{-7} watt/cm^2.

The disadvantage of this type of image convertor is that its front surface is restricted by the mechanical strength of the thin quartz disc which has to withstand a pressure differential of one atmosphere. Smyth[11.15] has suggested overcoming this difficulty by mounting a mosaic of quartz plates behind a metal plate able to withstand the pressure differential.

11.8 Ultrasonic Viscometers

We saw in section 9.5 that the reaction of a liquid on a torsional transducer can be related to the liquid's viscosity and hence this reaction can used to measure viscosity. Viscometers using torsional crystals are inconvenient for industrial use since they are fragile and not always suitable for use at high temperatures. Bradfield[11.16] has described one type using a tubular magnetostrictive torsional resonator made from a cobalt–iron–vanadium alloy with a Curie temperature of about 930° C which has operated satisfactorily so far at up to 180° C and 3250 atmospheres pressure.

References

11.1 Lamb, H., *Dynamical Theory of Sound*, 2nd edn., p. 122. Arnold (1925).
11.2 Hanstock, R. F., Lumb, R. F. and Walker, D. C. B., *Ultrasonics*, **2**, 109 (1964).

11.3   Rooney, J. and Reid, A., *Ultrasonics*, **4**, 57 (1966).

11.4   Dussik, K. T., *Z. phys. Med.*, **1**, 140 (1948).

11.5   Wild, J. J., *Surgery*, **47**, 183 (1950).

11.6   Howry, D. H., *I.R.E. Convention Record*, **3**, 75 (1955).

11.7   Howry, D. H., *Proc. 3rd Symposium on Ultrasound in Biology and Medicine*, University of Illinois (1964).

11.8   Baum, G. and Greenwood, I., *Proc. 3rd Int. Conf. on Med. Electronics, I.E.E.*, p. 412 (1963).

11.9   Donald, I. and Brown, T. G., ibid., p. 458.

11.10 Hopkin, P. R. and Haslett, R. W. G., *Ultrasonics*, **2**, 65 (1964).

11.11 Kay, L., *Ultrasonics*, **2**, 53 (1964).

11.12 Arenberg, D. L., *I.R.E. Convention Record* (1954).

11.13 Pohlman, R., *Z. Angew. Phys.*, **4**, 181 (1948).

11.14 Smyth, C. N., Poynton, F. Y. and Sayers, J. F., *Proc. I.E.E.*, **110**, 16 (1963).

11.15 Smyth, C. N., *Brit. Pat. No. 801294*.

11.16 Bradfield, G., *Proc. I.E.E.*, **109**, Part C, 565 (1962).

Appendices

Appendix 1

Piezoelectric Transducer Data

For an air-backed plate of piezoelectric material, of thickness l m and front surface area a m², vibrating in a thickness mode at its fundamental resonant frequency f_r under an applied voltage of V volts r.m.s. and radiating into a medium of specific acoustic impedance Z, we have:

$$\text{Transformer turns ratio} = hC_o = \left(\frac{2h\varepsilon}{c}\right) af_r$$

$$\text{Intensity output} \quad = \left(\frac{16h^2\varepsilon^2}{c^2}\right)\frac{f_r^2\,V^2}{Z}\ \text{Watt/m}^2.$$

$$\text{Radiation resistance} \quad = \left(\frac{c^2}{16h^2\varepsilon^2}\right)\frac{Z}{f_r^2\,a}\ \text{Ohms}$$

These terms, and other piezoelectric data, are tabulated below for quartz and some ceramic transducers.

Term	Quartz	Ceramic B*	PZT4*	PZT5H*
$h \times 10^{-8}$	43·2	16·7	26·8	21·5
$\varepsilon \times 10^9$	0·039	8·0	5·6	13·0
$\left(\dfrac{2h\varepsilon}{c}\right) \times 10^3$	0·059	4·87	7·5	14·0
$\left(\dfrac{16h^2\varepsilon^2}{c^2}\right) \times 10^4$	1·39 10⁻⁴	0·95	2·26	7·83
$\left(\dfrac{c^2}{16h^2\varepsilon^2}\right) \times 10^{-3}$	7·2 10⁴	10·5	4·43	1·28
$\rho_c c \times 10^{-7}$	1·43	3·04	3·0	3·0
Curie point, °C	576	115	328	193

Ceramic B is a slightly modified form of barium titanate.

* Trade-name of Brush Clevite.

Appendix 2

Ultrasonic Data for some Liquids

Liquid	f mc/s	$\alpha/f^2 \times 10^{15}$ sec^2/m	c m/sec	$\rho c \times 10^{-6}$	ref.
mercury	20–50	6·1	1500	19·8	1
ethyl iodide	15	40	869	1·68	2, 10
ethyl bromide	15	62	892	1·27	2, 9
methyl iodide	1–4	820	834	1·90	2, 10
	15	247			
methylene bromide	30	567	971	2·38	3, 9
methylene chloride	30	1114	1092	1·46	3, 10
chlorobenzene	30–192	147	1291	1·43	4, 5, 10
nitrobenzene	1–192	74	1473	1·78	2, 5, 10
methyl alcohol	1–192	30	1123	0·89	5, 10
benzyl alcohol	1–192	79	1540	1·61	5, 10
carbon tetrachloride	1–100	533	938	1·50	2, 10
water	1–192	21	1497	1·49	5, 6, 11
transformer oil	1		1425	1·28	12
benzene	0·1	900	1326	1·16	2, 5, 10
	104	849			
	192	775			
acetic acid	0·5	9.10^4	1144		5, 8
	67·5	158			
	192	139			
toluene	0·15	205	1328	1·15	5, 7, 10
	104–192	85			

References

1. Riekmann, P., *Physik Zeits.*, **40**, 582 (1939).
2. Pellam, J. R. and Galt, J. K., *J. Chem. Phys.*, **14**, 608 (1946).
3. Sette, D., *J. Chem. Phys.*, **19**, 1337 (1951).
4. Sette, D., *Nuovo Cimento* (9), Suppl. **2**, 7, 318 (1950).
5. Heasell, E. L. and Lamb, J., *Proc. Phys. Soc.*, **77**, 870 (1960).
6. Greenspan, M. and Tschiegg, C. E., *J. Res. Natl. Bur. Standards*, **59**, 249 (1957).
7. Biquard, P., *C.R. Acad. Sci. Paris*, **206**, 897 (1938).
8. Lamb, J. and Pinkerton, J. M. M., *Proc. Roy. Soc.*, A, **199**, 114 (1949).
9. Parthasarathy, S., *Proc. Indian Acad. Sci.*, A, **3**, 519 (1936).
10. Schaaffs, W., *Zeits. phys. Chem.*, **194**, 28 (1944).
11. Seifer, N., *Zeits. phys. Chem.*, **108**, 681 (1938).
12. Pancholy, M., Pande, A. and Parthasarathy, S., *J. Sci. and Ind. Res.*, **3**, 5 (1944).

Appendix 3

Ultrasonic Velocity and Impedance Data for some Solids

Solid	Velocity m/s		$\rho c \mid$ long. $\times 10^{-6}$
	Longitudinal	Transverse	
aluminium	6260	3080	16·9
copper	4700	2260	41·8
brass	4430	2123	36·1
nickel	5630	2960	49·5
crown glass	5660	3420	14·1
perspex	2670	1121	3·2
polystyrene	2350	1120	2·3
rubber	1479		1·4

Appendix 4

Selected Bibliography

1. Wavemotion and propagation

Stephens, R. W. B. and Bate, A. E., *Acoustics and Vibrational Physics*, Arnold (1966).
Morse, P. M., *Vibration and Sound*, McGraw-Hill (1948).
Redwood, M., *Mechanical Waveguides*, Pergamon (1960)—a thorough mathematical treatment of the propagation of elastic waves in bounded media.

2. General ultrasonics

Bergmann, L., *Der Ultraschall*, Hirzel Verlag, Stuttgart (1954)—an encyclopaedic study of the elementary theory and applications of ultrasonics with an extensive literature survey. A literature survey covering the period 1954–57 was published as an appendix in 1957.
Hueter, T. F. and Bolt, R. H., *Sonics*, Wiley (1955).
Blitz, J., *Fundamentals of Ultrasonics*, Butterworths (1963).
Handbuch der Physik, Vol. 11, Springer Verlag (1962).
Mason, W. P. (ed.), *Physical Acoustics*, Vols. 1, 2, 3, 4, Academic Press (1964)—these volumes are the most detailed and authoritative survey of the theory and application of ultrasonics (excluding industrial applications) which has so far appeared.

3. Absorption and dispersion of ultrasonic waves

Sette, D. (ed.), *Dispersion and Absorption of Sound by Molecular Processes*, Academic Press (1964).
Nozdrev, V. F., *The Use of Ultrasonics in Molecular Physics*, Pergamon (1965).
Herzfeld, K. F. and Litovitz, T. A., *Absorption and Dispersion of Ultrasonic Waves*, Academic Press (1959)—a thorough treatment of the subject although limited to gases and liquids only.
Mason, W. P., *Physical Acoustics and the Properties of Solids*, Van Nostrand (1958).

4. Piezoelectricity

Cady, W. G., *Piezoelectricity*, McGraw-Hill (1946)—the standard work on the subject.
Mason, W. P., *Piezoelectric Crystals and their Application to Ultrasonics*, Van Nostrand (1949).

5. Industrial applications

Crawford, A. E., *Ultrasonic Engineering*, Butterworths (1955).
Brown, B. and Goodman, J. E., *High Intensity Ultrasonics—Industrial Applications*, Iliffe (1965).

Babikov, O. I., *Ultrasonics and its Industrial Applications*, Consultants Bureau (1960).
Tucker, D. G. and Gazey, B. K., *Applied Underwater Acoustics*, Pergamon (1966)—concerned mainly with sonar and is good on transducer applications.

6. Medical and biological applications

Gordon, D. (ed.), *Ultrasound as a Diagnostic and Surgical Tool*, E. & S. Livingstone (1964).
Went, J. M. van, *Ultrasonic and Ultrashort Waves in Medicine*, Elsevier (1954).
Brown, B. (ed.), *Ultrasonics in Biology and Medicine*, Iliffe (1967).

7. Journals

Journal of the Acoustical Society of America.
Acustica, Hirzel Verlag.
Ultrasonics, Iliffe.

Principal Symbols Used

In general only those symbols which are used in more than one section are included in this list. The numbers in brackets refer to the section or chapter in which the symbol is defined or used extensively. Quantities defined *ad hoc* are not listed.

a	area of radiating surface of piston (chap. 3)
c	ultrasonic wave velocity
	magnetostriction constant (chap. 4)
c_0	ultrasonic velocity as frequency tends to zero (chap. 6)
c_∞	ultrasonic velocity as frequency tends to infinity (chap. 6)
c_g	group velocity (1.18)
c_p	phase velocity (1.14)
c_{rs}	elastic coefficients (1.3)
d_{ij}	piezoelectric strain coefficient (3.3)
e_{ij}	piezoelectric stress coefficient (3.3)
	strain component (1.3)
f	frequency
g_{ij}	piezoelectric strain coefficient (3.3)
h_{ij}, h	piezoelectric stress coefficient (3.3)
j	$\sqrt{-1}$
k	complex propagation constant (10.6)
	bulk modulus of elasticity
l	acoustic path length (1.12, 1.16)
	transducer thickness (chap. 3)
m	impedance ratio (1.15)
p	ultrasonic pressure (often used with suffices)
t	time
u	particle velocity (1.6)
x, y, z	rectangular co-ordinates
B	magnetic flux density (chap. 4)
C_p	specific heat at constant pressure (chap. 6)
C_v	specific heat at constant volume (chap. 6)
C'	relaxing specific heat (chap. 6)
$\bar{C}_p$	molecular specific heat at constant pressure (chap. 7)
$\bar{C}_v$	molecular specific heat at constant volume (chap. 7)
$\bar{C}'$	relaxing molecular specific heat (chap. 7)
C_{Py}	specific heat at constant pressure of component y (chap. 6)
C_o	static capacitance of transducer (chap. 3)
D	electrical displacement (chap. 3)
E	Young's modulus (chap. 1)
	ultrasonic energy density (1.7)
	electrical field strength (chap. 3)
G', G''	components of complex shear modulus (9.3)

I	ultrasonic intensity (1.7)
K	compressibility (chap. 6)
K_h	thermal conductivity (chap. 6)
N	number of turns of magnetostrictive transducer winding (chap. 4)
P	ultrasonic pressure amplitude (1.6)
P_a	static fluid pressure (chap. 5)
P_{rad}	radiation pressure (2.7)
Q	quality factor (3.10)
	charge on transducer plane surface (chap. 3)
	gas pressure within cavity at maximum radius (chap. 5)
Q_{mech}	mechanical Q of transducer (3.10)
R	intensity reflectivity (1.16)
	cavity radius (chap. 5)
	gas constant per mole (chap. 7)
R_m	maximum cavity radius (chap. 5)
	quantity defined in sections 6.5 and 6.8
R_p	equivalent parallel electrical resistance of transducer (3.9)
R_s	equivalent series electrical resistance of transducer (3.7)
S	strain (chap. 3)
	surface tension (chap. 5)
	entropy (chap. 6)
	cross sectional area of magnetostrictive transducer (chap. 4)
T	intensity transmissivity (1.16)
	tension (chap. 3)
	temperature
U	particle velocity amplitude
V	volume (chap. 6)
W	ultrasonic power (often used with suffices)
X_p	equivalent parallel electrical reactance of transducer (3.9)
X_s	equivalent series electrical reactance of transducer (3.7)
Z	impedance
	ratio defined in section 5.11
Z_{sp}	specific acoustic impedance (1.10)
Z_o	specific acoustic impedance of medium forming transmission line (1.12)
Z_c	specific acoustic impedance of transducer medium (chap. 3)
α	absorption coefficient for ultrasonic waves
α_{class}	classical value of absorption coefficient
β	term defined in section 1.6
γ	ratio of specific heats
	gyromagnetic ratio (10.7)
ε_{ij}	absolute permittivity (chap. 3)
η	fluid viscosity
	transducer efficiency (3.8)
λ	Lamé's first constant (chap. 1)
	magnetostriction constant (chap. 4)
	wavelength
μ	Lamé's second constant (chap. 1)
	shear modulus, rigidity (chap. 1)
	absolute permeability (chap. 4)

ν	Poisson's ratio (chap. 1)
ξ	displacement
ρ	density
ρ_c	density of transducer medium (chap. 3)
σ_{ij}	stress component (chap. 1)
τ	relaxation time (6·4)
τ', τ'', τ'''	relaxation times defined in section 6.6
ω	angular frequency (1.6)

Subject Index